PERFORMANCE INDICATORS FOR WATER SUPPLY SERVICES

BENCHMARKING AND PERFORMANCE ASSESSMENT SPECIALIST GROUP

BENCHMARKING AND PERFORMANCE ASSESSMENT

the international water association

PERFORMANCE INDICATORS FOR WATER SUPPLY SERVICES

THIRD EDITION

Helena Alegre, Jaime Melo Baptista, Enrique Cabrera Jr., Francisco Cubillo, Patrícia Duarte, Wolfram Hirner, Wolf Merkel, Renato Parena

PUBLISHING

Published by **IWA Publishing**
Alliance House
12 Caxton Street
London SW1H 0QS, UK
Telephone: +44 (0)20 7654 5500
Fax: +44 (0)20 7654 5555
Email: publications@iwap.co.uk
Web: www.iwapublishing.com

First published 2017

British Library Cataloguing in Publication Data
A CIP catalogue record for this book is available from the British Library

ISBN: 9781780406329 (Hardback)
ISBN: 9781780406336 (eBook)

TABLE OF CONTENTS

LIST OF FIGURES

LIST OF TABLES

The sigma software containing the full IWA performance indicators system, can be downloaded free of charge at http://www.sigmalite.com/iwamanual/

FOREWORD TO THE THIRD EDITION

This third edition is published almost two decades after the start of IWA's performance indicators journey. This was a long and intensive learning process. In the beginning, we all doubted that it would be feasible to establish a standardised system of performance indicators applicable to different contexts. The conclusion of the launching brainstorming session, in 1997, was that it would be a challenging and risky task but worth a trial. The first edition, published in 2000, showed that it is indeed possible and desirable to adopt standardised performance indicators. Nowadays, this doubt no longer exists. The international field test IWA carried out between 2000 and 2003 was essential to provide maturity to the initial PI system and learn about its applicability. The IWA PI systems started to be recognised as "THE" international standard to assess performance of the urban water services. Implementations of a much different nature started to appear worldwide.

The second edition of this manual, as well as the first edition of its twin manual focusing on wastewater services, greatly benefited from the experience achieved. This second version was born with one goal in mind: improving the existing tool, making it more powerful and useful. Experience showed that users tended to select too many indicators, lacking a structured process. Some early users argued that the system was too complicated in appearance and either too large (users could not use all 133 indicators) or too limited (users could not find the indicators they needed in the set). The answers to these questions were clear for the authors (not all indicators need to be used on every occasion, and additional indicators can be added when needed) but apparently, they were not so clear in that first manual.

The establishment of guidelines for assessing the drinking water and wastewater services in the international standards ISO 24510/24511/24512, on "Activities relating to drinking water and wastewater services", was another important step forward. Published in 2007 and fully aligned with the IWA PI systems, recognised as the main reference in this context, these standards emphasise the value of objective-oriented performance indicators to assist management. This standard greatly contributed to spread the message that the design of any performance assessment system starts by establishing the objectives of the activity and of the related assessment criteria. The performance indicators – or other types of metrics that may be more appropriate, shall fit them.

The IWA PI systems are nowadays in the core of performance assessment systems all over the world, in many different contexts. The manual has been published in multiple languages: Chinese, French, Persian, Portuguese and German, with partial translations in other languages, such as Czech, Japanese and Spanish. The proceedings of the IWA biannual PIxx conferences, starting in 2007, allow to trace evolution.

With the level of impact and maturity of the second edition, **why is a third edition needed**? Is it going to bring major changes to the PI system or to the key principles previously recommended?

> This third edition reflects the accumulated experience and recent learning, bringing a fresh set of inspiring cases that show how the IWA PI system has been successfully applied as a management tool, as a regulatory and public policies instrument and as the support for benchmarking initiatives.

The third edition keeps the core of the previous version. The PI system is the same, with some adjustments, particularly regarding financial definitions. Weighting the pros and cons of introducing other changes (e.g. eliminating indicators not frequently used or introducing some new ones), the conclusion was that the disadvantages of changing were considerable, with no major advantages. The cases show that customisation is easy and often desirable. This new version highlights the importance of the process of building a performance assessment system, objective-oriented. It also shows that PIs can peacefully coexist with other types of metrics (e.g. performance indices and levels, risk metrics), when these are more adequate to address the assessment criteria.

Helena Alegre, Leader – IWA Working Group on Performance indicators

Enrique Cabrera Jr., Chair – IWA Benchmarking and Performance Assessment Specialist Group.

ACKNOWLEDGEMENTS

As leader of the IWA Working Group on "Performance indicators for water supply services", I wish to express my appreciation to all those who made this work possible.

It is a great honour to realise that the full listing of contributors to this IWA manual of best practice would perhaps be as long as the manual itself. This is why it is not feasible to acknowledge everybody individually. In fact, we received contributions and support from every corner of the world, from the Far East (Japan) to the West (Brazil), from the Northern regions of Norway and Canada to the Southern regions of South Africa and Argentina, taking benefit from the true international dimension of IWA. On behalf of all the authors, I would like to register my deep appreciation to all individuals and organisations that made this manual happen.

A general acknowledgment is due to the more than 200 contributors of the First Edition. They played a key role in the preparation of the structure and contents of the IWA performance indicators systems.

The second phase of the Task Force work was directed to the field test of the performance indicators system and relied on the contributions from the 70 volunteer *utilities* that participated in the test. They provided support and enthusiasm, identified shortcomings and solutions to overcome them, assisted in the process of system debugging, helped setting up a methodology for the implementation of a performance indicator system and provided suggestions for the contents of the current Second Edition of the Manual. The wealth of information came primarily from the variety of participant types (bulk and direct suppliers, water only and multi-utilities, holding companies and regulators), levels of development (developed and developing regions participated), sizes (from less than 10 thousand to more than 20 million people supplied), organisational framework, and regions of the world (Europe, Asia-Pacific, Africa and South America). The National projects set up in Portugal (lead by the National Civil Engineering Laboratory and with the participation of 17 *utilities*) and in Germany (lead by the IWW Rhenish-Westphalia Institute for Water Research and with the participation of 14 German and one Swiss *utility*) played a key role, as they allowed for a direct contact between the co-ordinating team and the *utilities* testing the system. Also the workshops organised in Valencia (Spain, 2001) and Catania (Italy, 2002) for the field-test participants were a major source of relevant information allowing for the consolidation of the work. Complementary workshops organised by the Task Force team in Maputo (Mozambique, 2002) and Melbourne (IWA Congress 2002, Australia) are milestones to acknowledge as well.

The First Edition of the Manual was totally or partially translated into Portuguese, French, German, Spanish, Czech, Italian, Arabic and Japanese, on a volunteer basis and from the initiative of the individual countries involved. This helped to spread the word ensuring the international acceptance of the system. Conversely, the process of translation brought to light the lack of clarity of some concepts in the original version and helped to fix them. The Portuguese, Spanish and German translations were taken care of by authors of this Manual. The extended German version, published in 2005 and based on the Revised 2nd Edition, maintained the international structure while adapting specific German conditions and extensions of the methodology. The French translation was made on the basis of interactive work with the Task Force core team which enhanced the quality of the original.

In parallel with the field-test of the IWA performance indicators system, a sister IWA Task Force prepared the first edition of the IWA Manual of Best Practice for Wastewater Systems (published by IWA Publishing in 2003). Both activities were devolped in close link and with mutual inputs and benefits. Very warm thanks are due to our dear "dirty water" colleagues.

The authors also gratefully acknowledge all the organisations who sponsored the project, and in particular the International Water Association, the ITA group from the Polytechnic University of Valencia (ITA, Spain) and the National Laboratory of Civil Engineering (LNEC, Portugal).

IWA provided institutional support, encouragement and "seed money" to make the field test possible.

ITA also provided manpower for the project, and in particular for the preparation of the software application SIGMA.

Last but not least, I express my appreciation to my organisation, the National Civil Engineering Laboratory (LNEC), in Lisbon (Portugal), for allowing and encouraging me to complete this project and for having supported it financially. The LNEC fully supported financially the thousands of hours during our working schedule which Jaime, Patrícia and I spent on this project.

Helena Alegre

Leader of the IWA Working Group on Performance indicators

THE AUTHORS

Helena Alegre

Laboratório Nacional de Engenharia Civil (LNEC) – DH – NES, Av. do Brasil, 101

PT-1700-066 Lisboa, Portugal

Telephone (direct): +351 21 844 3625; E-mail: halegre@lnec.pt

Born in Castro Verde, Portugal, in 1957, she graduated in civil engineering from Lisbon University, Portugal in 1980, and obtained her doctorate degree in 1992 in civil engineering hydraulics with the thesis "Decision support tools for technical management of water distribution systems". She has been Senior Research Officer of the National Laboratory of Civil Engineering (LNEC) since 1994 and was Head of the Urban Water Division between 2002 and 2010. She served as Senior Vice-President of IWA in 2010–2012, is a member of the IWA Strategic Council, Chairs the IWA Strategic Asset Management Specialist Group (SAM SG), and integrates the Management Committee of the Benchmarking and Performance Assessment SG. In total, she (co)-organised more than 50 international and 60 national technical and scientific meetings. She is the author and co-author of about 400 publications, including 11 books. Her professional activity has always been related to urban water services and infrastructure asset management (IAM): applied research and consultancy in analysis, diagnosis, design and decision support for water supply systems, including performance assessment, strategic planning, system rehabilitation and water losses management.

Jaime Melo Baptista

Instituto Regulador de Águas e Resíduos (IRAR), Centro Empresarial Torres de Lisboa, Rua Tomás da Fonseca, Torre G, 8.° andar

1600-209 LISBOA, PORTUGAL

Telefone: +351 210 052 200; Email: Jaime.baptista@irar.pt

He graduated in civil engineering from Oporto University, Portugal, in 1975, and specialised in sanitary engineering at the New University of Lisbon in 1976. He became Researcher at the National Laboratory of Civil Engineering (LNEC) with the thesis "Economical design of water distribution systems" in 1983. He has been Principal Research Officer since 1992. He was the Head of the Water

Supply and Wastewater Division of LNEC between 1984 and 1990 and Head of the Hydraulics Department between 1990 and 2000, co-ordinating a team of 120 people dealing with applied research on water. He is the author and co-author of more than 500 scientific and technical publications and has organised a large number of national and international technical and scientific events. His field of expertise is public policy, regulation, rehabilitation, modelling, and performance assessment of water services, and has developed about twenty large research projects for the water industry. He Chaired the Portuguese Water Supply and Wastewater Standardisation Committee between 1990 and 1994. He is actively involved at the international level, particularly through the International Water Association (IWA). He has been a member of the IWA board of directors since 2012. He was honoured with the IWA Award for Outstanding Contribution to Water Management and Science. Between 2003 and 2015 he has been the president of the Portuguese Regulatory Authority of Water and Waste Services (ERSAR), an independent administrative authority, regulating about 500 utilities providing public water supply, urban waste water and urban waste management services in Portugal. He is presently Principal Research Officer of LNEC and Chairman of the Strategic Council of the Portuguese Water Partnership (PPA).

Enrique Cabrera Jr.

ITA, Universitat Politecnica de Valencia

Edificio 5C – Camino de Vera s/n – Valencia – Spain

Tel: +34 96 387 98 98; Fax: +34 96 387 98 99; E-mail: qcabrera@ita.upv.es

Enrique Cabrera is Associate Professor of fluid mechanics at the Universitat Politecnica de Valencia, Spain. Also a co-author of the IWA Manual of Best Practice on Benchmarking Water Services, Enrique has been working for over fifteen years on the topic of performance assessment and benchmarking.

Enrique plays an active role in IWA; he is a Director of the Association, Chair of IWA Publishing and Chair of the IWA Specialist Group on Benchmarking and Performance Assessment. Previously, he acted as secretary of the Efficient Operation and Management of Urban Water Systems and Planning and Construction Specialist Groups for 10 years. In 2008, he received the IWA Young Water Professional Award.

His international experience includes his role as a convenor of one of the working groups drafting of the ISO 24500 standards on the assessment and improvement of water services. In addition, Enrique has been part of the team that developed AquaRating, the standard developed by the Inter-American Development Bank and IWA to provide a rating for urban water and wastewater services, and chairs its Technical Committee. He has also acted on advisory and consulting roles in performance assessment projects worldwide, and published several books and numerous papers and journal articles on the topic.

Enrique holds two master degrees in industrial engineering and efficient use and management of water, and a PhD with a thesis on Performance Assessment and Benchmarking of water services by the Universitat Politecnica de Valencia.

Francisco Cubillo

Canal de Isabel II Gestión S.A., Calle José Abascal 10

28003 Madrid – Spain

Tel.: +34 91 545 1102; E-mail: fcubillo@canalgestion.es

Born in Chinchón, Madrid (Spain), he graduated in civil engineering from the School of Public Works Engineering, Madrid Polytechnic University, 1972. He is the Deputy Director of Research, Development & Innovation in Canal de Isabel II Gestión S.A., the company responsible for supplying, distributing and treating water for more than 6 million inhabitants in more than 170 municipalities, in the Madrid region. In IWA, he chairs the cluster of Alternative Resources. Between 2001 and 2011 he chaired EO&M (Efficient Operation and Management) Specialists Group and heads the International Conferences Efficient. During the years 1995 to 1999 he chaired the Information Technology Committee in IWSA (International Water Supply Association). In AEAS, (Spanish Association of Water and Wastewater Suppliers) he is a member of the Permanent Committee Coordination Organ and Chairs the Research Development and Innovation Committee. He has more than 30 years of experience in the Management of Hydraulic Supply, Resources, Environment & Demand Management, gathered on the Madrid Region Government, within the Environment Agency and, previously, within the Water Resources Directorate. Before this time, he worked for several private engineering consulting companies. Francisco Cubillo has published 13 books and more than 100 technical papers, being Professor in a wide variety of courses in Management Supply Systems, Hydrology, Technological Development and Environment.

Patrícia Duarte

Telephone: 351 966 177 194;

E-mail: p.duarte.sanina@gmail.com

Born in Lisbon in 1973, she graduated in environmental engineering in 1996, in the Faculty of Sciences and Technology of Universidade Nova de Lisboa, and concluded her PhD degree in Civil Engineering Hydraulics in 2010, in Instituto Superior Técnico, with her thesis work on water supply distribution networks design regarding technical performance optimisation.

She was a member of the IWA Task Force on "Performance Indicators for Wastewater Services" and co-authored the corresponding manual published by the IWA Publishing in 2003. She was also a member of the IWA performance indicators for water supply services field-testing coordinating team.

She contributed to the conception of the quality of service assessment system used by ERSAR, the Portuguese water, wastewater and solid waste services regulator. She helped translating the IWA Water Supply Manual of Best Practices (Alegre *et al.*, 2000) into Portuguese, as well as the recent ISO 5000, 55001 and 55002 standards on asset management.

For the last 6 years, independently of the described activity, Patrícia has been working for EPAL, the Lisbon water supply company, as an Asset Management technician.

Wolfram Hirner

Erlenstegenstraße 118B, 90491 Nürnberg, Germany

Telephone: +49 911 593161; E-mail: dr.hirner@web.de

Born in 1938 in Reutlingen, Germany, he graduated in 1965 as "Diplom Ingenieur" in mechanical and chemical engineering from the University of Stuttgart, Germany. Until 1973 he was a Scientific Assistant at the Institute of Chemical Engineering at the University of Stuttgart, and obtained his doctorate degree in 1972 with the thesis "Mass transfer in two-phase gas/liquid reactors". Between 1973 and 1977 he was an operating manager for industrial waste removal in Bavaria, South Germany. In 1978 he joined the company EWAG, Energy and Water Supply Company, Nuremberg, Germany.

Until his retirement in 1999 he was the managing director for natural gas and water supply with regional and local supply tasks. In 1991 he started chairing the German DVGW National Committees of "Water transmission and distribution" and "Further education of engineers". From 1993 he chaired the IWA Committee "Operations and Maintenance" within the IWA Distribution Division. He has published about 100 national and international papers, mainly on practice-oriented topics of water supply and management, preferably water distribution. He took part and presented papers in many technical and scientific meetings on water supply. He has played an important role within the German field-test and contributed with significant adaptations and extensions of the system.

Wolf Merkel

IWW Water Centre (IWW)

Moritzstr. 26, D-45476 Muelheim an der Ruhr – Germany

Telephone (direct): +49 (0)208 40303-100; E-mail: w.merkel@iww-online.de

Moritzstr. 26, D-45476 Muelheim an der Ruhr – Germany

Telephone (direct): +49 (0)208 40303-100; Fax: +49 (0)208 40303-82; E-mail: w.merkel@iww-online.de

Born in 1966 in Wiesbaden, Germany, he graduated in process engineering in 1985 from the University of Karlsruhe, Germany with a focus on water treatment technology. After a year as visiting scholar at Carnegie-Mellon-University working on contaminated site remediation, he worked on anaerobic wastewater treatment at the University of Stuttgart, Germany, finishing his PhD in 1997. Since 1998 he has been with IWW Rhenish-Westphalia Institute for Water Research (Muelheim, Germany) as a consultant and researcher in drinking water technology and water supply systems. Since 2002 he has been managing director of IWW. Wolf Merkel was leader of the German national project within the IWA Field test on performance indicators, integrating the viewpoints of the 14 participants into the international project. There are more than 70 technical and scientific papers and three books authored or co-authored by Wolf Merkel. He is a member of the working groups on performance assessment of DVGW and consultant for BDEW, the German water associations. Recently, he finished research projects on structural conditions for water suppliers and on the development of key performance indicators for German water supply.

Renato Parena

Chief Executive Officer

Società Canavesana Acque S.r.l., SMAT Group, Strada del Ghiaro snc, 10081 – Castellamonte, Italy

Past Chief Financial Officer

Società Metropolitana Acque Torino S.p.A., Corso XI Febbraio 14, 10152 – Torino, Italy

Telephone: +39 0124 093105; E-mail: amministratore.delegato@sca-acque.it; renatoparena@yahoo.it

Born in Torino, Italy, in 1945, he graduated in Economics from Turin University, Italy in 1969. In late 2099, he retired from the Società Metropolitana Acque Torino S.p.A., the water works of Torino, Italy where he was Chief Financial Officer. Currently he is Chief Executive Officer of the Società Canavesana Acque of Castellamonte, Italy, a small water system belonging to the SMAT Group that provides water services to 52 municipalities in the Turin province. He is an authorised auditor and a member of the Italian Association of Financial Analysts. He is author and co-author of many technical publications, papers and reports at national and international Congresses, meetings and conventions. Particularly involved in the Benchmarking context, he actively co-operated in editing the IWA Manual of Best Practice on "Process Benchmarking in the water industry". With over forty year's experience in the water sector, he managed financial and legal aspects of the SMAT Group M&A operations and particularly in 1997, as project leader, the transformation of Azienda Acque Metropolitane of Torino from a municipalised company into a publicly owned PLC, and in 2001 the merger with the set-aside company managing wastewater treatment plants. Inside the water sector he is a technical consultant of the Italian Federation of Water Services, an individual member of the American Water Works Association and an IWA Fellow. He plays an active role within the International Water Association, being currently Honorary Chairman of the "Statistics and Economics Specialist Group" and is a member of the IWA Strategic Council Sub Committee for Specialist Groups.

MEASUREMENT UNITS AND SYMBOLS

EUR	Euro
%	percentage
km	kilometre
km^2	square kilometre
kPa	kilopascal
kW	kilowatt
kWh	kilowatt × hour
l	litre
m	metre
m^2	square metre
m^3	cubic metre
°C	degree Celsius

ABBREVIATIONS

CTI	Contribution to investment
EBT	Earnings before taxes
EBIT	Earnings before interests and taxes
No.	number
OI	Operating income

OTHER CONVENTIONS

Terms written in the format *text* are defined in II-1.4.

PART I – FUNDAMENTALS OF PERFORMANCE ASSESSMENT

The sigma software containing the full IWA performance indicators system, can be downloaded free of charge at http://www.sigmalite.com/iwamanual/

1. INTRODUCTION

1.1 Balance of over 15 years of the IWA PI systems

The first manual of best practice on performance indicators for water supply services was born in 2000. Since then, the industry and its different stakeholders have shown a general acceptance of the need of the assessment of performance of water services and the use of performance indicators. As a result, the IWA framework has been referenced in many different regions, and its indicators applied to many different contexts.

The influence of the IWA PI systems turned them during this time in the de-facto standard of the industry for performance assessment. The inclusion of their concepts in national and international standards, their use by regulators of water services around the world and the the inspiration or even stronger presence in most of the active international performance assessment and benchmarking projects are proof of this. The IWA PI systems are often quoted as a key IWA achievement, a successful project if the success was to be measured by the number of initiatives in which the IWA PI manual is present somehow (either as a reference, source of insipiration or even at the foundation of the project).

However, a lot remains to be done and challenges still lie ahead in the way forward. While many projects include the IWA performance indicators in some form or the other, the actual framework and methodology are not as recognizable and have not been fully implemented outside a few exceptions. This is especially true around issues concerning confidence grading and data quality or the coherence of the management systems and their indicators (often not built around clear and strategic objectives). Additionally, the IWA performance indicators do not fully satisfy the needs of developing regions.

Finally, performance assessment was also a key part of the Millenium Development Goals related to water and will play at least an equal part of the Sustainable Development Goals. Although the IWA performance indicators do not focus on the same macro level, there is still room for improvement in the metrics used in these initiatives, especially given the great importance the indicators have in the actual outcome.

1.2 The water industry context, drivers and challenges

There has been a significant evolution in the water industry regarding performance assessment since the publication of the first edition of this manual. The industry has witnessed an increasing maturity in the management of water services in many parts of the world. This has resulted in institutional reforms aiming at creating and adequate scale for the water services, increasing their efficiency, transparency and accountability by means of regulatory systems. These reforms are still ongoing in some countries and others will surely follow in the near future.

Additionally, full cost recovery principles have become a more common ingredient in the governance of water services, with the consequence of a modified perspective in terms of pricing and efficiency assessment. Environmental issues, with an increased focus on energy, are also now perceived with an increased importance. However, service sustainability as a concept is still not as present as it should, and the long term outlook for infrastructure sustainability remains one of the key issues that is seldom properly addressed.

Public water supply remains an essential service for communities, part of the so called "services of general interest", being vital to general welfare, public health and the collective security of populations, as well as to economic activities and environmental preservation. Since the last edition of this manual, the United Nations recognized the Human Right to Water and Sanitation reinforcing the notion that all individuals are entitled to sufficient, safe, acceptable, physically accessible and affordable water for personal and domestic uses. Water supply services are the main vehicle to guarantee this right around the globe.

Such services allow for major scale economies and the installation and production unit costs decrease as demand increases, to a given extent. On the other hand, there is also significant scope for economy, where the production unit costs tend to decrease as similar activities are integrated (e.g., where the water supply and the wastewater treatment and disposal are managed together), due to the synergies in the management of human resources, equipment and facilities.

High value assets, representing an intensive capital sector, also characterise such activities. Additionally, these are long term assets, built up over many decades, designed for peak situations and generating idle capabilities which are not used for long periods. The infrastructures show a high degree of immobilization, as they are intended for a specific purpose and for this reason their transaction is difficult. This market shows a high relationship between the asset values and the revenues generated and low demand-price elasticity, as they are essential structural services.

Different relevant markets can be identified in the water sector. The most important is the transaction market between water *utilities* and final users, which is a natural monopoly. Other markets are the transaction market between bulk and retail utilities, which is in general a legal monopoly, the transaction market between utilities and services and goods suppliers, which is a competitive market, and the transaction market with owners of water resources, which is a quasi-monopoly.

Such features contribute to limit competition within the sector with regard to the service to the consumer. In practical terms, only the existence of one operator for every served geographic area is feasible, which originates a monopoly, either at a local or regional level, the user in such a context being not able to choose either an operator or the most convenient price-quality ratio. Natural monopolies occur whenever the cost structure is characterised by the falling of the production average marginal costs as the productive system develops, due to the existence of scale economies. In such cases, the whole production cost for a given demand issue is inferior whenever the service supplier is not just one.

Of course such a fact does not induce a continuous incentive to the *water utility* to increase effectiveness and efficiency, and this important weakness of this sector must be minimised as much as possible.

As always, the water supply sector faces major challenges in the future, and these are in constant evolution. Current pressures on the sector include:

- the adaptation to climate change, taking into account the potential water stress and extreme events situations resultin from in;
- the reduction in energy use in an energy-intensive activity such as water supply;

- keeping pace with the population growth, especially in rapidly-growing urban areas, but also with social changes (migrations, aging, etc.);
- financial pressures induced by the global economy;
- the evolution in the users' expectations and the political pressures for tariff reductions;
- the increasing age of infrastructure and the overall sustainability of the service

Maintaining a truly sustainable system in the water supply sector is as difficult and important as making the initial capital investment. The provision of safe water is a service and requires a service-orientated attitude on the part of the water *utilities* involved. Water services may normally be set at an affordable level for the consumers and managed and operated in accordance with the principles of good business practice and with the regulations, which are intended to protect the consumers and the environment.

The form of management will vary according to the local situation, and can be public, private or a partnership between both. In any case the continuous improvement of the quality of service is an essential target to be achieved.

1.3 The IWA approach to water services management

The management of water services needs to follow an integrated approach. Historically, the technical management of the services has focused on the engineering side of the business, considering decisions from the top management and financial decisions as inputs or constraints. However, water services above other sectors, greatly depend on technical criteria to face strategic decisions. A critical portion of all investments in water services are destined to expand, maintain and renew assets (in network based systems); at the same time, the capacity to tackle new constraints (environmental, social, quality of service, etc.) is strongly dependent on network characteristcs. For example, the possible strategic decisions of expanding the network, increasing resilience to water shortages, improving the quality of water or reducing the energy use cannot be efficiently made independently of technical decisions. A performance assessment system should then be based on taking the pulse of the critical aspects driving the success of such initiatives.

For this reason, the IWA PI framework advocates for a selection of indicators that is directly based on the strategic objectives for the service. An integration that should be clearly established for strategic asset management.

Asset management should always have as an overarching objective the sustainability of the service itself with time. This implies understanding the evolution of assets, but also of demand and constraints (e.g., the need to improve quality of service, more demanding environmental standards or new challenges like climate change). Once the strategic, long-term objectives have been set, the medium and short-term objectives can follow, and the adequate performance assessment metrics chosen to control and help to achieve such objectives.

1.4 The increasing importance of performance assessment systems

Performance assessment systems have become a more common practice for water services worldwide in the past 15 years, but the need for their use has also increased in importance. Water services have always been framed in a complex reality, where macroeconomic, social and environmental drivers were relevant to the core functions of the utility. The amount of intertwining factors keeps growing and, therefore, so does the complexity of the services (e.g., the water-energy nexus has become much more relevant in the recent past, and direct and indirect greenhouse emissions are now part of the equation when managing water supply services).

This ever increasing complexity requires the use of systematic tools that provide a global view and support decision making. Tools that, like performance assessment systems, can be configured to simultaneously consider the different constraints and drivers in a simplified way.

Performance assessment systems can also be used to facilitate the communication among the different stakeholders involved in water supply services. By providing simple, aggregated, reliable information, performance indicators represent a good communication tool between the different management levels, between the utility management and the elected politicians and regulators and even with society at large.

Further examples on why performance assessment systems represent an increasingly important tool are:

- The pressure from society for increased transparency and accountability, as described by the Lisbon Charter.
- The increase of data available to utility managers (data flooding) from new information systems, real time data-acquisition sensors, tools and software
- Increased financial stress with little or no room for poor investments and inefficiencies in the use of capital
- Further requirements for cost recovery and the need to improve internal processes to gain efficiencies in the process.

1.5 Users of performance assessment systems for water supply services

The most important *entities* or *'stakeholders'* in water supply are:

- the *water utilities*, which can be public, private or combined organisations who manage the water supply systems;
- the *consumers* or *direct users*, with whom a *water utility* has a relationship between supplier and customer;
- the *indirect stakeholders*, who do not have a direct connection to the system but may be affected by it or its impact on the surrounding environment (e.g., the consequences of water mains bursts, the quantitative or qualitative impact of water abstraction, etc.);
- the *pro-active stakeholders*, such as environmental organisations, consumer protection agencies and other pressure groups;
- the *policy-making bodies*, at local, regional and national level;
- the *regulatory agencies*, responsible for economical and quality of service regulation and for setting up and verifying compliance with statutory and other obligations;
- the *financing agencies*, of particular importance in such a capital-intensive sector as the water industry.

Other interested parties may include international agencies, from humanitarian associations to political organisations and multi-national companies.

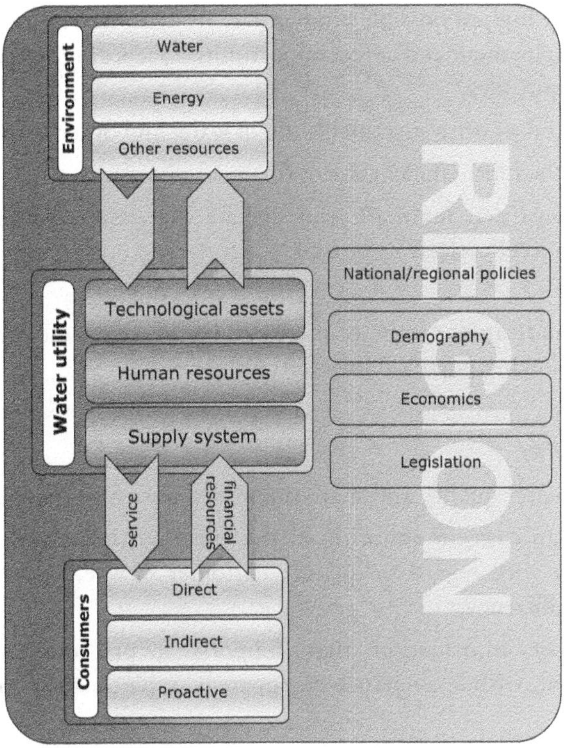

Figure 1. Water utility context.

Irrespective of their nature (private, public or combined) or geographical extent, it may be presumed that all water supply *utilities* share a common purpose and management objectives, which may be stated as: *"The achievement of the highest level of consumer satisfaction and service quality in line with the prevailing regulatory framework, whilst making best use of available resources"* (Faria & Alegre, 1996).

By taking into consideration the relationships between stakeholders, resources and values (see Figure 1), it is possible to identify five types of management objectives:

- to provide an appropriate level of service to consumers, while complying with national and regional policies and meeting statutory and other obligations;

- to obtain the highest possible productivity from human resources, and offer the best employment and career opportunities according to the individuals' skills and aptitudes;
- to protect and assure a sustainable use of water and other natural resources;
- to achieve the most efficient use of financial resources;
- to plan, construct, maintain and operate the *utility*'s physical assets as efficiently and effectively as possible.

Performance indicators can be of use to all of the above mentioned entities, having the following potential benefits and uses:

For *water utilities*:

- facilitates better quality and more timely response from managers;
- allows for an easier monitoring of the effects of management decisions, particularly with regard to quality, customer service, sustainability, and economic efficiency;
- provides key information that supports a pro-active approach to management, with less reliance on apparent system malfunctions (reactive approach);
- highlights strengths and weaknesses of departments, identifying the need for corrective measures to improve productivity, procedures and routines;
- assists with implementation of a Total Quality Management regime, as a way of emphasising all-round quality and efficiency throughout the organisation;
- facilitates the implementation of benchmarking routines, both internally, for comparing the performance at different locations or systems, and externally, for comparison with other similar *utilities*, thus promoting performance improvements;
- provides a sound technical basis for auditing the organisation's workings and predicting the effect of any recommendations made as a result of an audit.

For the national or regional *policy-making bodies*:

- provides a common basis for comparing the performance of water *utilities* and identifying possible corrective measures;

- supports the formulation of policies for the water sector, within the integrated management of water resources, including resource allocations, investments, and the development of new regulating tools.

For *regulatory agencies*:

- provides key monitoring tools to help safeguard consumer interests in a monopoly service supplier situation, assessing the performance and benchmarking the *utilities*, and monitoring compliance with contracted goals.

For *financing bodies*:

- assistance in assessing investment priorities, project selection and follow-up.

For *consumers* and *pro-active stakeholders*:

- provides the means of translating complex processes into simple-to-understand information and of transmitting a measure of the quality of service provided.

For *supra-national organisations*:

- provides a very appropriate language for identifying the main asymmetries between regions of the world, their causes and evolution, thereby assisting in defining strategies.

1.6 Objectives of iwa manuals of best practice series

The main objective of this manual is to provide guidelines for the establishment of a management tool for water supply *utilities* based on the use of performance indicators.

Complementary objectives are to provide:

- a coherent framework of indicators that may allow for future comparisons between *utilities* in the scope of initiatives of comparative performance assessment;
- a coherent framework of indicators that may be adopted by regulatory agencies;

- a coherent framework of indicators that allow for progressively enlarging and consolidating the international statistics collected by the IWA.

The IWA PI System includes the performance indicators potentially pertinent at the top strategic management level of a *utility.* It aims to incorporate all the relevant aspects required to express management objectives and results in terms of an *utility*'s performance. Complementary indicators may be needed depending on local context and circumstances, and will be necessary when assessing at the function, process and tasks level. However, these tend to be much more organisation dependent. Therefore, they are not included in the system, but can easily be accommodated within the same structure and principles.

1.7 About this manual

The reader is invited to read sequentially the first part (Part I) of the manual. Starting with the basic concepts used, the text explains which are the key components of the IWA-PI system and its rationale and recommends a general implementation methodology. Some examples of application illustrate in very different cases how this system can be customised to the specific needs of each user without losing its characteristic of being an international reference. Part II, the "yellow pages", contains the IWA-PI system itself, with the definition of all system elements and their relationships. Whilst the first part aims to be a proper manual on how to implement and use performance indicators, the second part is the necessary reference tool to apply them.

2. SYSTEMS OF PERFORMANCE INDICATORS

2.1 Performance indicators system

The final goal of any performance indicators system is to provide information. It is important to make the distinction between information and data. A correct definition for information would be "data that can be used for the purpose of making decisions". Consequently, a system of performance indicators is not only aimed at providing the value of a few ratios, but also all the complementary elements (quality of the data, explanatory factors, context) which are needed in order to make appropriate decisions.

A performance indicators system is consequently the result of considering all areas of interest, stakeholders and influencing factors in a certain environment. In the case of water *utilities*, a comprehensive system would include the whole company, the stakeholders, the users, the environment, and all related areas that may be worth monitoring for management purposes.

As a consequence, a system of performance indicators comprises a set of performance indicators and related data elements which represent real instances of the *utility* context (Figure 2). The classification of these data elements depends on the active role they play:

Data elements: A basic datum from the system which can either be measured from the field or is easily obtainable. Depending on their nature and role within the system, data elements can be considered variables, context information or simply explanatory factors.

Variables: A variable is a data element from the system which can be combined by processing rules in order to define performance indicators. The complete variable consists of a value (resulting from a measurement or a record) expressed in a specific unit, and a confidence grade which indicates the quality of the data represented by the variable.

Performance indicators: Measures of the efficiency and effectiveness of the delivery of the services that result from the combination of several variables. The information provided by a performance indicator is the result of a comparison (to a target value, previous values of the same indicator, or values of the same indicator from other *utilities*).

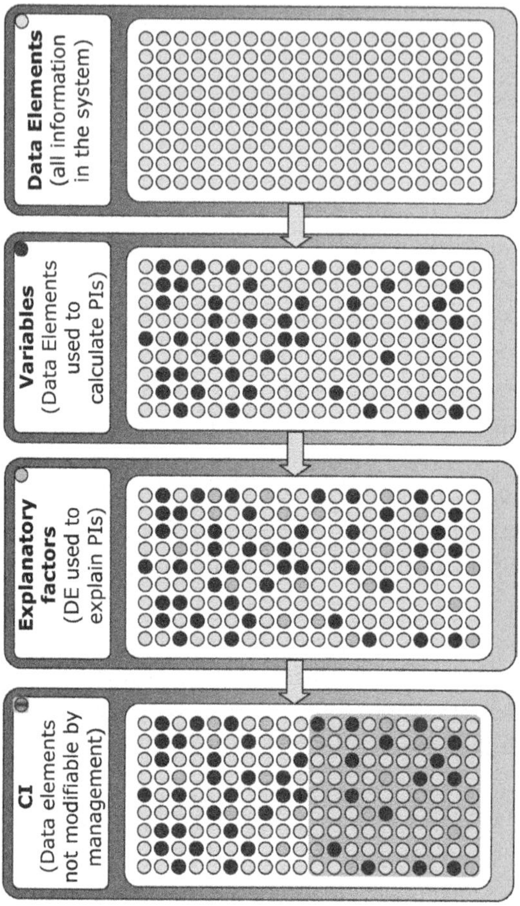

Figure 2. Components of a performance indicators system.

An individual PI should be unique and collectively appropriate for representing all the relevant aspects of an *utility*'s performance in a true and unbiased way, thus reflecting the managing activity. Each performance indicator should contribute to the expression of the level of actual performance achieved in a certain area and during a given period of time, allowing for a clear comparison with targeted objectives and simplifying an otherwise complex analysis.

A performance indicator consists of a value (resulting from the evaluation of the "processing rule") expressed in specific units, and a confidence grade which indicates the quality of the data represented by the indicator.

Performance Indicators are typically expressed as ratios between variables; these may be commensurate (e.g., %) or non-commensurate (e.g., $/m^3). In the latter case, the denominator shall represent one dimension of the system (e.g., number of service connections; total mains length; annual costs), to allow for comparisons. The use of denominators of variables which may vary substantially from one year to another, particularly if not under the control of the *utility*, should be avoided (e.g., annual consumption, that may be affected by weather or other external reasons), unless the numerator varies in the same proportion. A rule of thumb is that the denominator should be a cause of change in the numerator.

A clear processing rule should be defined for each indicator, specifying all the variables required and their algebraic combination.

Context information: Context information are data elements that provide information on the inherent characteristics of an *utility* and account for differences between systems. There are two possible types of context information:

- Information describing pure context and external factors to the management of the system. These data elements remain relatively constant through time (demographics, geographics, etc.) and in any case are not affected by management decisions. In some countries (e.g., Germany) this information is known as "structural information".
- Some data elements on the other hand are not modifiable by management decisions in the short or medium term, but the management policies can influence them in the long term (for instance the state of the infrastructure of the utility).

Context information is especially useful when comparing indicators from different systems.

Explanatory factors: An explanatory factor is any element of the system of performance indicators that can be used to explain PI values, i.e., the level of performance at the analysis stage. This includes PI, variables, context information and other data elements not playing an active role before the analysis stage.

The use of performance indicators should always be linked to the establishment of a proper performance assessment system, in which all the above mentioned elements are present and defined, and aim to fulfil a clear objective or obtain information on specific areas or issues. The proposal in this manual represents in itself a complete system of performance indicators that can be used on an "as is" basis, completed with further elements or simplified through a selection of part of its elements to suit particular needs.

2.2 Requirements for the definition of a system of performance indicators

A good number of problems which originate in the use of performance indicators can be solved at the definition stage of the PI system. Setting the objectives and constraints of the system is helpful when choosing and defining the indicators. Although the definition and selection of performance indicators is dealt with in the implementation chapter, there are a few principles that should be taken into account at the definition stage.

Performance indicators

Individually, a performance indicator should comply with the following requirements:

- be clearly defined, with a concise meaning;
- be reasonably achievable (which mainly depends on the related variables);
- be auditable;
- be as universal as possible and provide a measure which is independent from the particular conditions of the utility;
- be simple and easy to understand; and
- be quantifiable so as to provide an objective measurement of the service, avoiding any personal or subjective appraisal;
- include information on the data quality of the variables.

Collectively, PI should comply with the following requirements:

- every PI should provide information significantly different from the other PI in the system;
- definitions of the performance indicators should be unequivocal (this requirement is made extensive to its variables);
- only PI which are deemed essential for effective performance evaluation should be established.

PIs should be objective-oriented and comply with the commonly accepted standard concepts and principles, such as ISO 24510:2007, 24511:2007, and 24512:2007 and IAS/IFRS (International Accounting Standards/ International Financial Reporting Standards enacted by the International Accounting Standard Board – IASB) (*see II-1.3*). In economically regulated services the regulatory accounts should be taken into account.

Variables

Each variable should comply with the following requirements:

- definitions should be univocal;
- fit the definition of the PI they are used for;
- be reasonably achievable;
- refer to the same geographical area and the same period of time or reference date as the PI they will be used for;
- be as reliable and accurate as the decisions made based on them require;
- include information on data quality.

Some of the variables in PI systems are often obtained from external data, and their availability, accuracy, reference dates and limits of the corresponding geographical area are generally out of the control of the *utility*. In this case, variables should also comply with the following requirements:

- be collected, whenever possible, from official survey departments;
- be fundamental for the PI assessment or interpretation; and
- collectively, be as few as possible.

Context information and other data elements

Context information and the rest of the data elements in the system (which are used as explanatory factors) should follow the same general principles as variables and performance indicators. However, the level of detail and confidence grading is usually not considered to be as high as the one required for PI and variables. Consequently, CI and the rest of the data elements should comply with:

- definitions should be univocal;
- be reasonably achievable;
- if external, be collected whenever possible from official survey departments;
- be fundamental for the PI interpretation; and
- collectively, be as few as possible.

3. FROM PERFORMANCE ASSESSMENT TO IMPROVEMENT

3.1 Performance indicators as the foundation of benchmarking

In 2011, the IWA Manual of Best Practices "Benchmarking water services" (Cabrera Jr. *et al.* 2011) was published as a logical follow-up to the first editions of this manual. Being part of the same series, the contents of this chapter are fully consistent with those of the other manual and sometimes they are even coincident. In any case, for a full insight on benchmarking and its application to water services, the reader should refer to the original manual.

Although the term benchmarking has been used in many different ways in the water industry for almost two decades, the truth is that in all those conceptions of the technique, performance indicators were a key part of the process.

As explained previously, performance indicators can be used in different ways, but in essence they provide an excellent tool to assess the current status of an organization by providing a detailed snapshot of key performance areas.

This capacity has proven to be essential for benchmarking efforts. After all benchmarking requires identifying the best performer in one way or the other, and indicators are the perfect tools for the job.

It is important to point out that the terms "metric benchmarking" and "process benchmarking" should be avoided as recommended in the aforementioned manual:

> *"The IWA Specialist Group on Benchmarking strongly recommends abandoning the use of the terms "metric benchmarking" and "process benchmarking". Instead "performance assessment" and "performance improvement" should be considered consecutive components of benchmarking."*
>
> (Cabrera Jr. *et al.* 2011)

This may come as a surprise to readers of the second edition of this manual, where those terms were used to describe the different roles of performance indicators. However, although the concepts remain perfectly valid today, it is true that different actors in the industry often used the names a contradicting manner in the past, and the strong recommendation seeks to provide a common and unmistakable language.

Details and historical references on this issue can be found in the benchmarking manual, while the formal definition of benchmarking follows in this chapter.

In any case, and to summarize the recommendations of the IWA Specialist Group on Benchmarking and Performance Assessment, the use of the following terms is recommended:

- **Performance assessment:** Use of performance indicators to determine the current status and the evolution of the performance of a water or wastewater service.
- **Comparative performance assessment:** Use of performance indicators to determine the relative performance of a water or wastewater service with respect to some peers (formerly referred to as metric benchmarking).
- **Performance improvement:** Second phase of the benchmarking process in which best practices are identified and adapted to improve performance in a water or wastewater service.

All these tools (and their names) can be used at different levels of detail, as shown in Figure 3 (Cabrera Jr. *et al.* 2011)

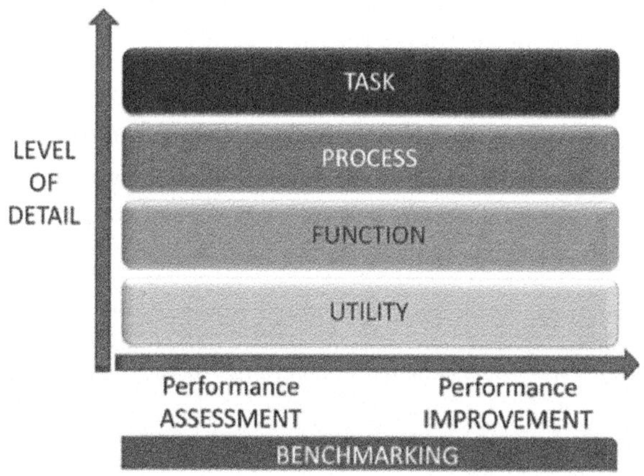

Figure 3. The performance assessment and improvement model.

3.2 What is benchmarking?

There are plenty of definitions of benchmarking in the literature. Most of those references are not addressed to the water sector, a fact that should have little or no influence in the adequacy of those definitions. However, until the publication of the IWA MBP on benchmarking (Cabrera Jr. *et al.* 2011) there was a good deal of contradicting definitions on what benchmarking is within the water industry.

The definition provided in the IWA manual of 2011 is a classical one, and encompasses in a single sentence the essence of benchmarking:

> "Benchmarking is a tool[1] for performance improvement through systematic search and adaptation of leading practices"

The careful choice of words aims to capture in a single sentence a broad concept. And key to that concept is the idea that benchmarking is simply a tool. A powerful tool that is especially suited for the water industry, but in no way an end in itself. Benchmarking without a clear goal will often lead to disappointment and waste of resources.

Another cornerstone of the benchmarking concept is captured by the word systematic. Benchmarking techniques should always be aimed at continuous improvement. As a matter of fact, benchmarking fits especially well in the PDCA (Plan, Do, Check, Act) concept and should always be approached with those four stages in mind. As a natural consequence, benchmarking should derive in any organization in a natural tendency towards continuous improvement.

However benchmarking is anything but a "do it yourself" concept. The search for leading practices implies that lessons need to be learnt from others[2], hopefully from the best in class. But as much as benchmarking is about looking outside, it is also an exercise of looking within and learning how things are done internally. It is only from that inner knowledge and the understanding of how others (the best) do things, that improvement can be achieved.

An obvious factor to that success is understanding the concepts of *best practices* and *best in class*. Those words are often quoted in the literature pointing

---

1 Benchmarking should probably be considered more a process than a tool. However, for reasons later explained, the authors have refrained from using the term "process" in this definition.

2 These others may certainly include other utilities from a same group of companies.

to iconic case studies and world-class companies as references in certain processes. While it is true that those organizations constitute a clear reference, benchmarking is not limited to industry leaders and large organizations. And as a matter of fact, some of the benchmarking projects undertaken lately by water *utilites* show that lessons can be learnt from almost anyone, and that being able to identify the best practices and those who have developed them is a key factor to success.

The truth is that to find out whom the best in class is, measurements need to be made. However, and despite the very extended notion that benchmarking, or some type of benchmarking, is achieved by comparing metrics, it should be made very clear that creating a bar graph with a numerical comparison of several *utilites* is not benchmarking. Even the original definition of "metric benchmarking" (Kingdom *et al.* 1996) considered it to be much more than the comparison of a few indicators. The benchmarking concept has always encompassed a systematic process and the will for continuous improvement.

The consequences of continuous benchmarking efforts in a *utility* often derive in a more mature organization which is more transparent and understands better how things are done, which things need to and could be improved and what it takes to improve them.

3.3 Why should you benchmark?

Performance indicators have been and will be used as an independent tool without the need to proceed onto benchmarking. The mere use of indicators (individually or in comparison with others) will not improve the performance of a *utility*. PIs are an excellent tool to assess performance and compare it to past values, targets or peers' performance, but solutions to possible shortcomings and improvement must be achieved with complementary tools. Benchmarking is one of such tools.

However, history shows that many projects that started as performance assessment projects have evolved into benchmarking as their participants required making use of the improvement opportunities that indicators showed. The 6 cities project in Scandinavia or the European Benchmarking Cooperation are a clear example of this.

In many occasions, the drive to benchmark was initiated outside the *utility*, and benchmarking was seen as a logical option after one of the following happened:

> Public debate about liberalisation or privatisation of public services; demand for more transparent and efficient public services; political pressure for full cost recovery or requirements for large investments to improve the service in terms of coverage and quality, which in their turn require smart innovations by the water industry to keep charges at a reasonable level.

However external drivers are usually not sufficient to initiate benchmarking (excluding regulatory obligations). Usually, the *utility* management, with its operational responsibility for the service, has its own drivers to benchmark. The natural need to continuously improve the organisation and its products can be facilitated in benchmarking projects that provide a detailed insight in the performance and identify areas and means of improvement. The relative position of a *utility* against its peers can help determining the sense of urgency for taking measures in the analysed areas.

However, if something has also been made clear during all these years is that many utilities are reluctant to enter the benchmarking arena. Therefore, we should also acknowledge that there may be reasons for a *utility* to refrain from joining a benchmarking program:

> The *utility* is considered to be unique and not comparable; the company is being restructured or merged, so there is no stable situation for assessing performance; there is a lack of reliable data to submit; there are not enough available resources (budget, manpower); there are doubts on the added value of the program; the suggested methodology is too complicated; there is no guarantee for confidentiality of individual performance data.

Therefore, these issues shall be kept in mind when a benchmarking activity is prepared.

The benchmarking scene is not limited to utilities. As a matter of fact, the industry itself has sometimes been less interested in benchmarking than other stakeholders.

Governments/regulators

Governments are politically responsible for water services. Regulators, when present, focus more on the individual *utility* level. In any case, comparative performance assessments can help both governments and regulators to introduce artificial competition in a sector that constitutes a natural monopoly and to put pressure on utilities to raise efficiency and transparency.

Govenments & regulators – Why assess performance?

- Political responsibility for the regulation of the water sector.
- Need to guarantee:
  - Appropriate levels of service and compliance with applicable standards.
  - Sustainable and efficient operations, transparency of the water industry.
  - Appropriate organisation/regulation of the water market (natural monopoly)
- Benefits from comparative performance assessments:
  - Insight in performance of the water sector (compliance, service levels, investments etc.)
  - Pressure on utilities for efficiency.

Figure 4. Governments and regulators: Why assess performance? (Cabrera Jr. *et al.* 2011).

Customers

Customers and consumer organizations are usually focused on obtaining a good service and a good product while paying a fair price. Additionally, questions regarding affordability for those with the lower incomes are often a main concern.

Benchmarking is an excellent tool for consumers, and so is comparative performance assessment (it is not difficult to find consumer driven reports comparing several utilities).

Customers – Why benchmark?

- Require adequate water services with appropriate level of service.
- Expect affordable prices and value for money.
- Benefits from comparative performance assessments:
  - Insight in the performance of the local utility.
  - Service provided by a utility continuously seeking efficiency and in competition with peers.
  - Comparative assessment of value for money.

Figure 5. Customers: Why benchmark? (Cabrera Jr. *et al.* 2011).

Owners/shareholders

The owners and shareholders are legally responsible for the *utility*. As a result of such responsibility, and to comply with applicable regulations, they need insight in the *utility*'s performance, its efficiency and the magnitude of the financial or other risks undertaken by the *utility*.

Benchmarking and comparative performance assessment are excellent tools to assess those magnitudes and place them in perspective when compared with others. Additionally, benchmarking demonstrates that there is a culture within the organization to continuously improve and become more efficient.

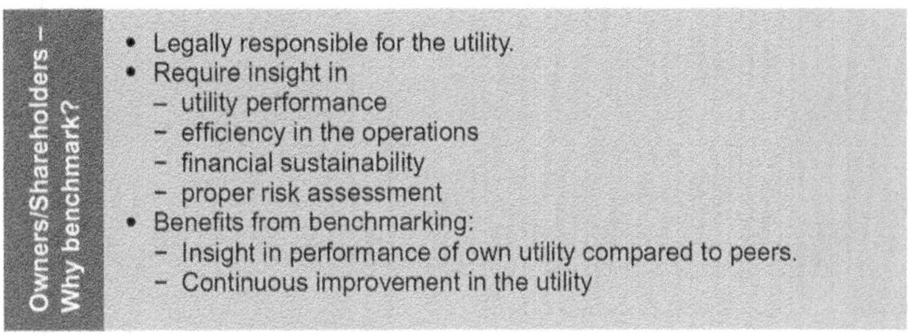

Figure 6. Owners/shareholders: Why benchmark? (Cabrera Jr. *et al.* 2011).

3.4 The IWA Benchmarking Framework

A quick look at the IWA Benchmarking framework (Figure 7) can provide an estimate of the complexity and work required to run a successful benchmarking project. This framework is not only relevant for those involved in such project, but also for organizations and individuals dealing with performance indicators projects.

The similarities between comparative performance assessment (comparison of PIs) and benchmarking projects are many. As a matter of fact, a benchmarking project will always include a mini-project in which performance is compared.

The IWA benchmarking framework comprises three stages:

[1] **Preparation:** The project is planned, establishing the objectives and the resources needed to develop it. The performance assessment system is designed and implemented, including data collection. The internal rules for the project are laid out.

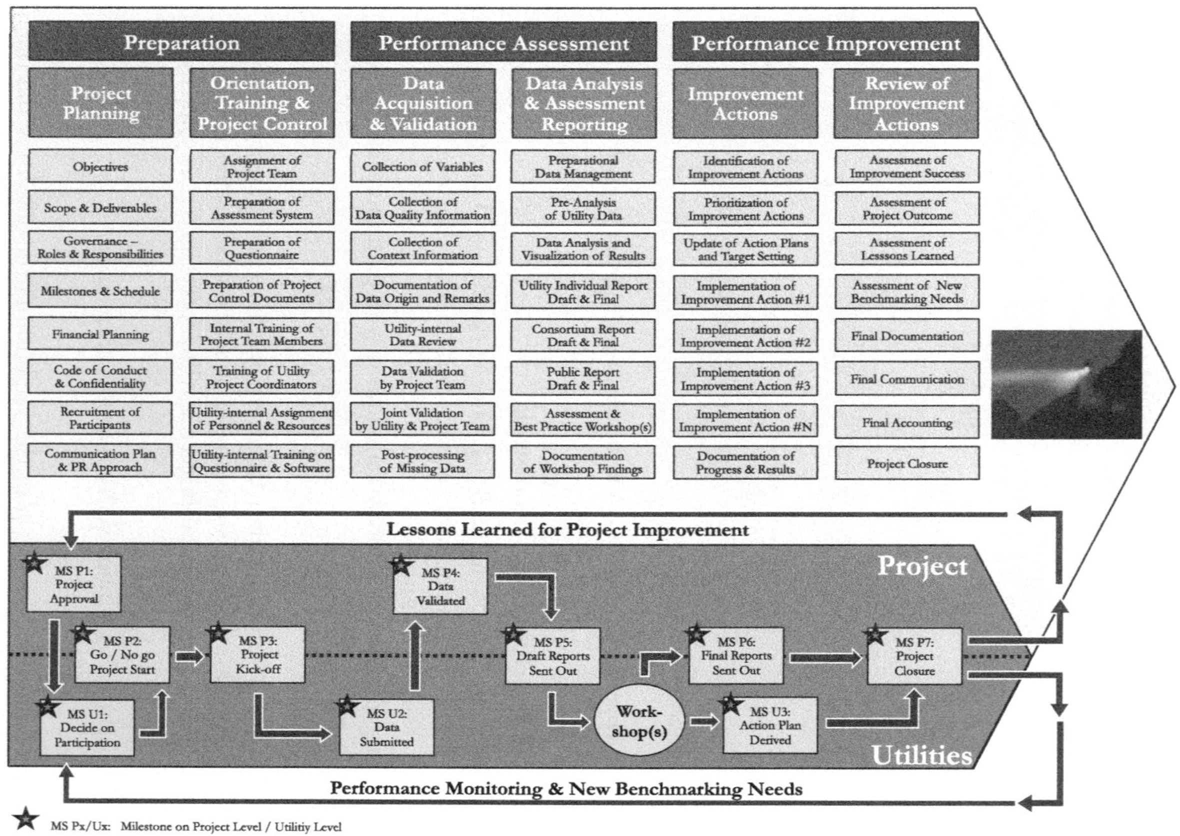

Figure 7. The IWA benchmarking framework (Cabrera Jr. *et al.* 2011).

[2] **Performance Assessment:** Variables, context information and data quality information are collected. Once data has been validated, performance indicators are calculated. Results are analysed and reports drafted.

[3] **Performance improvement:** Improvement actions are identified and achievable targets are set. Performance indicators are used to track the success of the project. Reports are created and new benchmarking needs assessed.

All performance indicators related projects may share many individual steps and the nature of some stages with the IWA benchmarking process. Table 1 shows the degree of correlation between three typical projects involving the use of performance indicators in relationship with the IWA benchmarking framework.

Table 1. Degree of correlation between the stages of the IWA benchmarking framework and typical performance indicators related projects.

	Preparation	Performance Assessment	Performance Improvement
Internal performance assessment	Medium	High	None
Comparative performance assessment	High	High	None
Benchmarking	High	High	High

4. FROM PERFORMANCE ASSESSMENT TO INFRASTRUCTURE ASSET MANAGEMENT

4.1 What is infrastructure asset management?

Infrastructure asset management (IAM) is the set of coordinated activities that an organization uses to realize value from the water supply system in the delivery of its outcomes or objectives. Realization of value requires the achievement of a balance of costs, risks and benefits over different timescales.

IAM aims to reply to questions such as:

- Who are we at present and what service do we deliver?
- What do we own in terms of infrastructures?
- Where do we want to be in the long term?
- How do we get action?
- How can we prove that our decisions are effectively addressing the strategic objective?
- How can we quantify the impact on the corporate objectives of our decisions and of subsequent actions?

IAM is most often approached based on partial views: e.g., for business managers and accountants, IAM means financial planning and the control of business risk exposure (Harlow & Young, 2001); for water engineers, IAM is focused on network analysis and design, master planning, construction, optimal operation and hydraulic reliability (Alegre & Almeida, 2009); for asset maintenance managers, the infrastructure is mostly an inventory of individual assets. IAM tends to be the sum of asset-by-asset plans, established based on condition and *criticality* assessment; for many elected officials, since water infrastructures are mostly buried, low visibility assets, IAM tends to be driven by service coverage, quality and affordability in the short run. Common misconceptions include reducing IAM to a one-size-fits-all set of principles and solutions, mistaking it for a piece of software, substituting it for engineering technology, or believing that it can be altogether outsourced. In practical terms, many existing implementations tend to be biased by one or several of these perspectives.

The IWA IAM framework allows for comprehensive and unbiased implementations. It requires the establishment of a sound objective-based performance assessment system. It is in accordance with the principles and requirements of the International Standards ISO 55000 – Asset Management.

4.2 Why should you implement an IAM approach?

IAM is increasingly becoming a key topic in the move towards compliance with performance requirements in water supply and wastewater systems. Sustainable management of these systems should respond to the need for:

- Promoting adequate levels of service and strengthening long-term service reliability;
- Improving the sustainable use of water and energy;
- Managing service risk, taking into account users' needs and risk acceptance;
- Extending service life of existing assets instead of building new, when feasible;
- Upholding and phasing in climate change adaptations;
- Improving investment and operational efficiency of the organisation;
- Justifying in a clear and straightforward manner the investment priorities.

Effective decision-making requires a comprehensive approach that ensures the desired performance at an acceptable risk level, taking into consideration the costs of building, operating, maintaining and disposing capital assets over their life cycles. Brown and Humphrey (2005) summarize these concepts by defining IAM as "the art of balancing performance, cost and risk in the long term".

4.3 The IWA IAM framework

The cube shown in Figure 8 demonstrates the key dimensions of the IWA IAM approach. It advocates that IAM must be addressed at different planning decisional levels: a *strategic level*, driven by corporate and long-term views and aimed at establishing and communicating strategic priorities to staff and citizens; a *tactical level*, where the intermediate managers in charge of the infrastructures need to select what the best medium-term intervention solutions are; and an *operational level*, where the short-term actions are planned and implemented. It also draws attention to the need for standardised procedures to assess intervention alternatives in terms of performance, risk and cost, over the analysis period. The other relevant message is that IAM requires three main pillars of competence: business management, engineering and information.

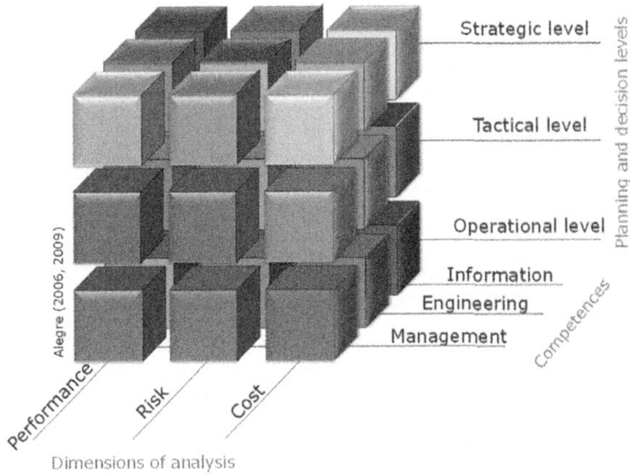

Figure 8. The recommended IAM approach (Alegre, 2009).

At each level of management and planning – strategic, tactical and operational –, a structured loop (Figure 9) is proposed that comprises the following stages: (i) definition of objectives and targets; (ii) diagnosis; (iii) plan production, including the identification, comparison and selection of alternative solutions; (iv) plan implementation; and (v) monitoring and review. Most utilities already have several elements of this process in place. What is often missing is a review mechanism – a way to measure compliance with set goals – as well as an effective alignment between the different management levels.

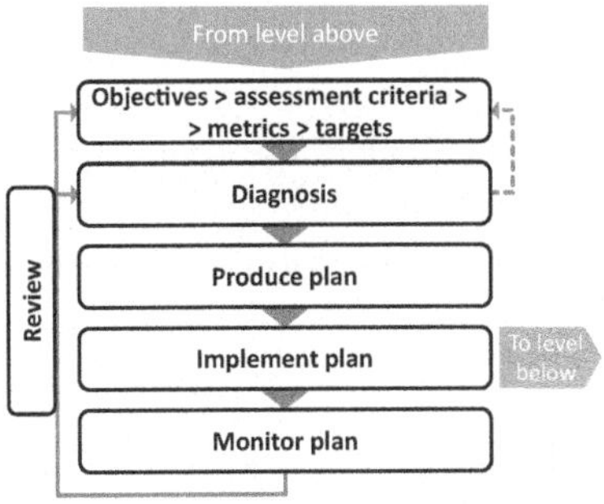

Figure 9. The planning process (at each planning level).

Setting up objectives, assessment criteria, metrics and targets is a crucial stage in order to set up clear directions of action, as well as accountability of results through timely review. Clarifying the four distinct but sequential concepts:

- *Objectives* are the goals that the organization aims to achieve. The IWA IAM approach, in accordance with the ISO 24510:2007, 24511:2007, 24512:2007 and EN 752:2008 standards, demands that objectives are clear and concise, as well as ambitious, feasible and compatible, and take into account the ultimate goal for the utility of providing a sustainable service to society. For each objective, it is recommended that key assessment criteria be specified.
- *Criteria* are points of view that allow for the assessment of the objectives. For each criterion, performance, risk and cost metrics must be selected in order for clear targets to be set, and for further monitoring of the results.
- *Metrics* are the specific parameters or functions used to quantitatively or qualitatively assess criteria; metrics can be indicators, indices or levels.
- *Reference values* are the judgement of what good, fair and poor for each metric and for the stakeholders across the board. This judgement shall be established independently from the specific cases and be as stable as possible over time.
- *Targets* are the actual proposed values to be achieved for each metric within a given time frame (short, medium or long term).

For instance: for an objective of *environmental sustainability*, one possible criterion could be *water usage efficiency*, measured through a *real losses per service connection (l/conn./day)* metric, for which a target of *100 l/conn./day* might be chosen.

Assessment metrics are a key element of the whole process: they are used to establish targets, to set up a diagnosis, to compare and select alternative courses of action, and to monitor and review the process. They should be relevant, reliable, simple, and effectively measure success. Objectives and targets are also a powerful means of communication within the organisation and with other stakeholders.

As illustrated in Figure 10, the process cascades through the decision levels within the organisation's management structure. The global approach is based on plan-do-check-act (PDCA) principles aiming at the continuous improvement of the IAM process. The key notions in this process are *alignment, feedback, involvement and empowerment*:

- **Alignment** among strategic, tactical and operational objectives and targets is paramount in ensuring that efforts and resources are not wasted in the long run. In other words, making the best out of limited resources cannot be achieved without smart alignment across the utility.
- **Feedback** is crucial because alignment cannot be assured through a top-down process alone. It is fundamental to have feedback mechanisms within each level, as well as between levels.
- People are the key element in this process. First of all, top management must be engaged in the shift of paradigm to an integrated IAM approach. It is equally important to ensure the involvement of the entire organization, from the CEO to the asset operators, and the **empowerment** of the staff, in order to promote **leadership**, co-ordination, collaboration, corporate culture acceptance, motivation, commitment and corporate know-how.

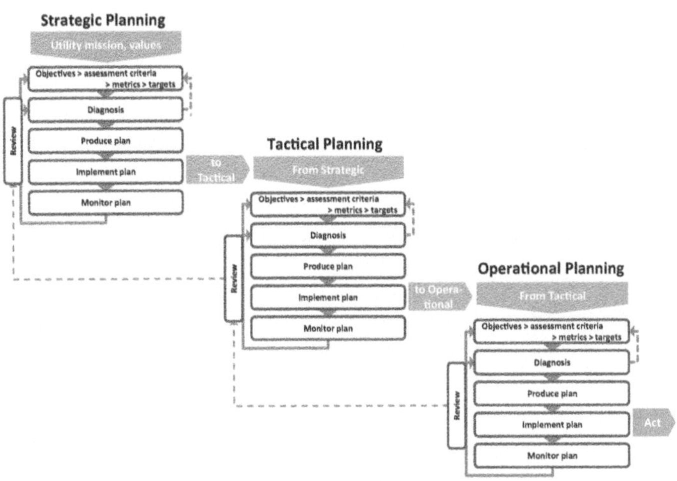

Figure 10. Alignment and feedback between planning (decision-making) levels.

5. THE IWA SYSTEM OF PERFORMANCE INDICATORS

5.1 Objective

In order to properly understand the IWA system of performance indicators in its conception, it is necessary to know the original objective of the work.

This manual originated as a result of the belief that performance indicators could be used in the water industry, and that some guidelines should be defined on how to define them and use them.

Anyone with a little experience in the field of performance indicators knows that there is almost an infinite number of possibilities in terms of selecting performance ratios and defining a system of performance indicators. Consequently this work must be understood as the result of applying a series of constraints, defining clear objectives and achieving consensus among the many people who have taken part in the project through the years. The main objectives that were originally pursued were:

- In order to provide a useful tool, the manual should not only contain the guidelines and theoretical principles to use indicators. Providing a robust, consistent and coherent set of performance indicators was established as a top priority and considered as a useful tool for the industry from the beginning.
- The system was to be as universal as possible. Water supply conditions around the world are varied and very different. In some cases, differences are so great that compatibility for some indicators is impossible. This has a reflection on the final system, which is redundant for some concepts with solutions for both ends of the spectrum (for instance, the concept of interruptions makes sense in countries where supply is continuous. In case of an intermittent supply, the frequency is what is really relevant).
- The system should be useful for all the stakeholders playing a role in the *water utility* context as described above. This obviously presented a wide range of users and different uses that had to be considered for the system. Although the use and final user of a PI system can certainly determine many aspects of the system itself, a common solution was the objective.
- Despite the fact that the system was developed taking into account the many players and uses in the water industry, it was mainly aimed at the industry itself. As a managerial tool, the main user in mind was always the utility itself,

and the goal, the improvement of the service. The field test stage that preceded this Second Edition was as a matter of fact mainly aimed at water companies of all sizes, nature and origin, in order to make the system more suitable for its natural users. Further widespread applications of the System proved that it may represent an excellent common and generally accepted tool to assess utility performance, suitable for international benchmarking projects.

- For obvious reasons, aspects that were regarded as particular to each *utility* were left out of the system. These included context characteristics such as geographical particularity, institutional framework or internal organization, but also parts of the business itself, such as water treatment.
- Last but not least, the number of performance indicators had to be limited. Through the years, the number of indicators and variation on the existing indicators probably reached over one thousand. Since the original set, the number of indicators has consistently been considered too large by almost everyone seeing it for the first time. However when most of these people were asked to make proposals for improvement, these often came in the shape of adding more indicators or substituting some, and hardly ever mentioned the removal or the decrease in the total number of PI.

5.2 Structure

Performance indicators

PI are grouped in a structure that makes sense for every utility and for all types of uses of the system. The performance indicators were arranged in the following groups:

- Water resources (WR)
- Personnel (Pe)
- Physical (Ph)
- Operational (Op)
- Quality of service (QS)
- Economic and financial (Fi)

These main groups are intended to help in identifying the purpose of a certain indicator and the final user within the *utility*. The indicator code, which is unique, consists of the two letter code shown above and a number which shows the order of the indicator within the group.

Each one of these groups is also divided into subgroups. Division into subgroups is also aimed at helping identify the use and user of a certain indicator or indicators (in some special cases, subgroups are also broken down into smaller units). For instance these are the subgroups included in the Operational group:

- Operational indicators
  - — Inspection and maintenance of physical assets
  - — Instrumentation calibration
  - — Electrical and signal transmission equipment inspection
  - — Vehicle availability
  - — Mains, valves and service connection rehabilitation
  - — Pumps rehabilitation
  - — Water losses
  - — Failures
  - — Water metering
  - — Water quality monitoring

Finally, some of the indicators are broken down into sub-indicators. Normally sub-indicators are parts of the top indicator which may or may not be assessed separately. For instance the "quality of supplied water" indicator which accounts for the number of successful quality tests with respect to the total can be broken down as follows:

- QS18 – Quality of supplied water
- QS19 – Aesthetic tests compliance
- QS20 – Microbiological tests compliance
- QS21 – Physical-chemical tests compliance
- QS22 – Radioactivity tests compliance

Variables

Variables are organised in the IWA PI system in a similar way to indicators. It is important to remember that variables are as a matter of fact the inputs of a PI system, and consequently they have to be either measured or obtained from real instances in the utility. The variable groups were established in accordance with the possible origin of the data, and regardless of the indicators for which they were used. As a matter of fact a variable can be used to calculate several indicators from different groups. The variables in the IWA PI system are divided into the following groups:

- A – Water volume data
- B – Personnel data
- C – Physical assets data
- D – Operational data
- E – Demography and customer data
- F – Quality of service data
- G – Financial data

Variables can sometimes also be broken down into parts, and consequently calculated using other variables. For instance, it may be convenient to consider the total pipe length for every material used in the *utility* (steel, ductile iron, polyethylene, etc.) although some indicators will only use the total length of pipes (which can be calculated by adding up all the partial lengths).

Attending to this consideration, it is possible to make a classification between *primary* and *secondary* variables. Primary variables are those directly involved in the calculation of the selected performance indicators. Secondary variables are used to calculate primary variables. Consequently this classification is not permanent and depends on the selected indicators. However, it is useful when determining the total number of data items needed to evaluate the performance indicators. It is also important to outline that primary variables are also measurable like any other variable, and although they can be calculated through secondary variables this is just an alternative.

Context information and other system data

Context information is a crucial part of the IWA PI system. However, it has an enormous dependency on the local conditions of the *utility* under study. A data item that is relevant to understand the context and its influence on the performance of an *utility* is negligible in another *utility.*

Consequently, the CI elements provided with the IWA system should be considered for reference only and a starting point in order to assess all possible explanatory factors due to context in a system.

The Context Information elements provided in the IWA PI system are classified in the following categories:

- Service data
- Physical assets
  - Water resources
  - Impounding reservoir storage
  - Treatment
  - Transmission and distribution storage tanks/service
  - Pumping stations
  - Transmission and distribution network
  - Service connections
- Consumption and peak factors
- Demography and economics
- Environment

5.3 Using the IWA system of performance indicators

Contrary to the general belief, the IWA system of performance indicators should be considered as a very flexible and adaptable tool. After the First Edition of the manual was published, many users expressed the view that the number of indicators was too high, the definitions did not serve their purposes or the time frame was not appropriate for their business. However, it never was in the authors' minds to impose those criteria on users' worldwide, but simply to turn the long process into an easier and shorter one.

Although the indicators, variables and context information in the IWA system have been chosen to try and represent universal concepts that may be applied almost anywhere in the world, the authors are well aware that they may be insufficient, incomplete or inappropriate in certain particular situations. It is not possible to define a system of PI that will cover all needs, points of view and circumstances. However, it is possible to define a starting point as broad and general as possible that will enable users to develop their own compatible systems.

These are some of the common misconceptions regarding the IWA system that should not influence its use:

- **The IWA system has too many performance indicators**: As explained later in the implementation chapter, one of the first steps to follow when using PI is to select the relevant ones. The IWA system should be viewed as a large supermarket where users can shop for "quality" indicators (with thorough definitions, consistent in their nature, and with the added value of potentially being used by many others). In any case, practical applications of the IWA system have already witnessed in the past years the exclusive use of IWA indicators, mixed solutions with own-defined indicators or solutions that only apply the IWA structure and none of its indicators.
- **The IWA system is too complicated to use**: On its own, the IWA system is not any more complex or simple than any other PI system. What typically characterises the IWA system is the coherent structure of its elements, a thorough definition of these elements and a complete set of support tools and literature. If stripped from all these additional elements the IWA system could be considered as just the structure (based on PI, variables, CI, and other data elements) and a consistent set of sample elements (PI, variables and CI) that can be used to configure a system for any goal within urban water supply.
- **The IWA system is too general and my *utility* has very specific needs**: As mentioned before, the IWA system cannot cover all particular cases and the conditions of every *utility* in the world. However, it is hardly imaginable that any *utility* in the world will not find useful at least a few of the IWA proposed indicators. The PI system for any *utility* should be defined in accordance with the needs and local conditions of such a system. The IWA proposal represents an excellent starting point to develop such a specific system. The same can be said about developing PI systems for the department level.
- **The IWA system is only evaluated yearly and I need feedback more often**: The evaluation period quoted in the elements of the IWA system is only there for reference. Depending on the management needs of every *utility*, the use of performance indicators will change. It may need yearly evaluations up to daily tracking depending on the activity and the indicators. Changing the evaluation period is another option permitted to provide flexibility to the system.
- **If the IWA system is so general, there is no real advantage in using it**: Using the concepts developed for the IWA system guarantees that the system will be consistent and coherent. Furthermore, the system elements

will be compatible with those of other systems in terms of nature and definitions (thus enabling the use of the developed software platforms, the integration in combined benchmarking sets, etc.).

- **I already have my information systems and do not want to use any other software**: Data management in the IWA system can be greatly improved by the use of computing tools. SIGMA is just an example tool with educational purposes and is not intended to provide a professional assessment of performance. On the other hand SIGMA is not by any means necessary to use the IWA PI system and every project should consider the need for data management tools separately.

5.4 Performance indicators

Underlying assumptions

The performance indicators contained in the IWA-PI system have been carefully selected in order to be as general as possible, keeping in mind the requirements listed in I-2.2. This chapter lists and describes them, referring the motivation for their selection. When appropriate, it also explains the reason for the inclusion of a given indicator or exclusion of aspects relevant for the management but for which it is not feasible to define meaningful performance indicators that fit the requirements.

All the indicators refer to the core business of water supply service. Multi-service utilities should assess the part of resources and revenues related to the water supply component. The same applies to *utilities* with complementary businesses, such as for instance, selling services related to water quality sampling and testing or repair of private property plumbing.

The full definitions of every indicator can be found in II-2, including processing rules and variable specification. For ease of use, the number of the starting page where details can be found is indicated between brackets in the title of the tables presented in this section.

The current version of the PI set is basically the same as the first one. The structure, the groups and the indicators themselves have not changed in their essence. The main criticism received was the inadequacy of the initial system to deal with bulk supply systems, and this shortcoming has been overcome.

However, the precise definitions have received a number of minor changes and refinements. Conversely, a small number of indicators were removed because they did not prove to be feasible or useful, and some new ones were added. Therefore, it is strongly recommended that all those who adopted indicators from the First Edition of the manual and did not participate in the field test, spend some time with the corresponding PI in this Edition and make themselves aware of the differences.

Let us now make a guided tour through every PI group.

Water resources indicators

The type and characteristics of the water resources vary greatly from case to case, both in terms of quantity and quality. The water resources indicators included in the IWA-PI system are limited to the common aspects of interest, i.e., how efficiently are the resources being used and is there a comfortable margin between water resources available and used. It should be noted that particular instances are likely to require other complementary indicators.

The IWA Water Resources PI are as follows:

Table 2. Water resources PI (II #202)[3].

WR1 –	Inefficiency of use of water resources (%)
	Percentage of water that enters the system and is lost by leakage and overflows up to the point of customer metering.
WR2 –	Water resources availability (%)
	Percentage of available water that enters the system.
	WR3 – Own water resources availability (%)
	Percentage of own resources available water that enters the system.
WR4 –	Reused supplied water (%)
	Percentage of water entering the system that is reused.

WR1 is the environmental indicator of water losses. It provides a measure of the water environmental inefficiency, and the field testers confirmed its relevance as such. However, it should be noted that it is neither appropriate to assess the infrastructure condition nor the economic water losses. The operational

[3] Legend: (II #151) = (See detailed information in Part II, page 151).

or economic water losses indicators should be used for these. The field-test demonstrated that WR2 is considered important by the *utilities*, but frequently difficult to assess. WR3 is new compared with the First Edition and was recommended by the field test participants. It provides an idea of dependency of the *utility* on third party water sources, when there is a small or medium share of imported water. However, when the external water resources dependency tends to 100%, the value of the indicator tends to infinity (∞) and becomes meaningless. WR4 is also a recommendation from the field test participants and is relevant for the *utilities* that incorporate reused water in the supply.

Personnel indicators

The personnel indicators address the viewpoints of efficiency of human resources (total personnel, personnel per main function and technical services personnel activities), qualification and training, health and safety, and overtime work.

The personnel indicators should be assessed in terms of full-time equivalent employees and their interpretation should be completed analysing in parallel the corresponding level of outsourcing.

The number of employees required to run a system greatly depends on the characteristics of the system itself. However, it also greatly depends on the performance of the management and therefore the IWA-PI system includes PI related to the total personnel, as a rather aggregated indicator, and more detailed indicators allowing for a more in-depth analysis of how the staff is distributed. The comparison of these type of indicators between *utilities* of similar characteristics help identify areas where there is room for improvement.

Total personnel PI are:

Table 3. Total personnel PI (II #204).

Pe1 –	Employees per connection (No./1000 connections)
	Number of full time equivalent employees per 1000 service connections.
Pe2 –	Employees per water produced (No./(10^6 m^3/year))
	Number of full time equivalent employees per unit volume of water produced.

Pe1 and Pe2 should be used alternatively. Pe1 is more appropriate for distribution systems and Pe2 for bulk supply systems.

The following group of PI allows for identifying the weight of manpower (in terms of number of employees) in each main function:

Table 4. Personnel per main function PI (II #205).

Pe3 –	General management personnel (%)	
	Percentage of employees dedicated to directorate, central administration, strategic planning, public relations, other stakeholder relations, legal affairs, internal audits, environmental management, new business development and general computing support.	
Pe4 –	Human resources management personnel (%)	
	Percentage of employees dedicated to personnel administration, education and training, occupational safety and health services and social activities.	
Pe5 –	Financial and commercial personnel (%)	
	Percentage of employees dedicated to economic and financial planning, economic administration, economic controlling and purchasing and material management.	
Pe6 –	Customer service personnel (%)	
	Percentage of employees dedicated to accounting and control and to customer relations and management activities.	
Pe7 –	Technical services personnel (%)	
	Percentage of employees dedicated to planning, construction, operations and maintenance activities	
	Pe8 – Planning & construction personnel (%)	
	Percentage of employees of technical services working in planning & construction.	
	Pe9 – Operations & maintenance personnel (%)	
	Percentage of employees of technical services working in operations & maintenance.	

Considering that the technical staff deals with the core of the business, and in general corresponds to the biggest share among all main functions, the following PI allow for a deeper analysis, identifying the personnel allocated to each technical activity. In this case, the indicators are expressed in terms of a quantity directly related to the type of activity.

Table 5. Technical services personnel per activity PI (II #207).

Pe10 –	Water resources and catchment management personnel (No./(10^6 m^3/year))
	Number of employees working in the water resources and catchment management per unit volume of water produced.
Pe11 –	Abstraction and treatment personnel (No./(10^6 m^3/year))
	Number of employees working in planning, design, construction, operations and maintenance of the abstraction and treatment per volume of water produced.
Pe12 –	Transmission, storage and distribution personnel (No./100 km)
	Number of employees working in planning, design, construction, operations and maintenance of the transmission, storage and distribution system per mains length.
Pe13 –	Water quality monitoring personnel (No./(10000 tests/year))
	Number of employees working in water quality sampling and testing per unit amount of tests made annually.
Pe14 –	Meter management personnel (No./1000 meters)
	Number of employees working in meter management per 1000 meters.
Pe15 –	Support services personnel (%)
	Percentage of employees working in support services.

An important management objective is to have an adequate balance of personnel qualification, assessed by the following indicators:

Table 6. Personnel qualification PI (II #209).

Pe16 –	University degree personnel (%)
	Percentage of employees with university degree.
Pe17 –	Basic education personnel (%)
	Percentage of employees with basic education.
Pe18 –	Other qualification personnel (%)
	Percentage of employees without basic education.

At least as important as the academic qualification is the professional qualification. However, the analysis of the criteria used to define qualifications categories showed that there are major differences from country to country that prevent the establishment of definitions of general international use. All the alternatives analysed proved to cause misleading interpretations and were eliminated. This problem can be overcome at the company or even country level. Whenever feasible and relevant, the above indicators should be complemented with (or replaced by) professional qualification indicators.

One of the management objectives is to keep well trained personnel. The IWA-PI system includes the following indicators to assess this point of view:

Table 7. Personnel training PI (II #210).

Pe19 –	Total training (hours/employee/year)
	Number of training hours per employee and year.
	Pe20 – Internal training (hours/employee/year)
	Number of hours of internal training per employee and year.
	Pe21 – External training (hours/employee/year)
	Number of hours of external training per employee and year.

The following PI are included in the IWA-PI system to address the viewpoint of health and safety:

Table 8. Personnel health and safety PI (II #211).

Pe22 –	Working accidents (No./100 employees/year)
	Number of working accidents requiring medical care per hundred employees and year.
Pe23 –	Absenteeism (days/employee/year)
	Number of days of absenteeism per employee and year.
	Pe24 – Absenteeism due to working accidents or illness at work (days/employee/year)
	Number of days of absenteeism due to working accidents or illness at work per employee and year.
	Pe25 – Absenteeism due to other reasons (days/employee/year)
	Number of days of absenteeism neither due to working accidents nor illness per employee and year.

Unplanned tasks and under-staffing cause the need for overtime work. This is therefore an important view point to consider:

Table 9. Overtime work PI (II #213).

Pe26 –	Overtime work (%)
	Overtime work expressed as a percentage of the normal working time.

Physical indicators

The physical indicators aim to address the aspects of asset performance in terms of capacity, related to their use. The main type of components are considered: treatment, storage, pumping and transmission and distribution pipes. Although of a different nature, the automation assets are also included as to the importance of their role.

The indicator related to treatment aims to assess the existing buffer in terms of treatment capacity. The interpretation needs to take into account the uncertainty of the demand and the ease of selecting alternative sources of treated water if a high demand occurs. Obviously, interpretation must consider the actual redundancy in case of several plants, delivering water to every point within the supply zone.

Table 10. Water treatment PI (II #213).

Ph1 –	Treatment plant utilisation (%)
	Maximum percentage of the daily capacity of the existing treatment plants that were used.

Two indicators are considered related to storage capacity. One refers to raw water storage (impounding reservoirs) and the other to treated water storage (transmission and distribution storage tanks).

Table 11. Water storage PI (II #214).

Ph2 –	Raw water storage capacity (days)
	Capacity of raw water reservoirs per unit volume of daily system water input.
Ph3 –	Treated water storage capacity (days)
	Capacity of transmission and distribution service reservoirs per unit volume of system water input.

Similarly to the case of treatment and storage assets, in the case of pumping one indicator refers to the existing buffer in terms of pumping capacity. The other ones relate to efficiency of energy use. The traditional indicator "energy consumption per m^3" is not included because it does not allow for comparisons between systems, although it may be useful to track the evolution of a given system. The indicator "energy recovery" is relevant when there are gravity flows with significant energy surplus that can be recovered.

Table 12. Pumping PI (II #215).

Ph4 –	Pumping utilisation (%)
	Percentage of the maximum pumping capacity (that can be used simultaneously) that was actually used.
Ph5 –	Standardised energy consumption (kWh/m^3/100 m)
	Average pumping energy consumption in the system per 1 m^3 at 100 m of head.
Ph6 –	Reactive energy consumption (%)
	Percentage of the total energy consumption for pumping that corresponds to reactive energy consumption.

Table 12.a. Pumping PI (II #158) (contd.).

Ph7 –	Energy recovery (%)
	Percentage of the total energy consumption for pumping that is recovered by the use of turbines of reverse pumps.

With regard to the transmission and distribution network, no capacity indicators are considered because no measure was encountered that fits the PI requirements established (be clearly definable, with a concise meaning, be reasonably achievable, be auditable; be simple and easy to understand; and be quantifiable so as to provide an objective measurement of the service, avoiding any personal or subjective appraisal). The indicators considered refer to availability of valves, hydrants and meters, essential components of the system for a proper operation.

Table 13. Valve, hydrant and meter availability PI (II #216).

Transmission and distribution	
Ph8 –	Valve density (No./km)
	Number of isolating valves per unit of distribution mains length.
Ph9 –	Hydrant density (No./km)
	Number of hydrants per unit of distribution mains length.
Meters	
	Ph10 – District meter density (No./1000 service connections)
	Number of district meters per thousand of service connections.
	Ph11 – Customer meter density (No./service connection)
	Number of direct customer meters per service connection.
	Ph12 – Metered customers (No./customer)
	Number of direct and bulk customer meters per customer
	Ph13 – Metered residential customers (No./customer)
	Number of residential-equivalent customer meters per residential customer.

One of the management objectives may be to increase the automation and remote control level. Two indicators refer to this matter:

Table 14. Automation and control PI (II #218).

Ph14 –	Automation degree (%)
	Percentage of control units that are automated.
Ph15 –	Remote control degree (%)
	Percentage of control units that are remotely controlled.

Operational indicators

Much of the efficiency of a *water utility* can be lost or improved in the operation and maintenance activities. Managers need to monitor the planned activities for inspection, preventative maintenance and rehabilitation, and the unplanned activities caused by the failure of system elements.

With regard to the inspection and maintenance of physical assets, the following PI are available in the IWA-PI system:

Table 15. Inspection and maintenance PI (II #219).

Op1 –	Pump inspection (-/year)
	Rate of pumps inspected per year, assessed in terms of nominal power.
Op2 –	Storage tank cleaning (-/year)
	Rate of storage tank cells cleaned per year, assessed in terms of volume.
Op3 –	Network inspection (%/year)
	Percentage of network inspected per year, assessed in terms of length.
Op4 –	Leakage control (%/year)
	Percentage of mains length subject to active leakage control.
Op5 –	Active leakage control repairs (No./100 km/year)
	Number of leaks detected and repaired due to active leakage control per unit of mains length.
Op6 –	Hydrant inspection (-/year)
	Rate of hydrants inspected per year.

Reliable monitoring information is based on accurate data. Instrumentation calibration is a key aspect:

Table 16. Instrumentation calibration PI (II #222).

Op7 –	System flow meters calibration (-/year)
	Rate of system flow meter calibrations per year.
Op8 –	Customer meter replacement (-/year)
	Rate of flow meter replacement per year.
Op9 –	Pressure meters calibration (-/year)
	Rate of pressure meter calibration per year.
Op10 –	Water level meters calibration (-/year)
	Rate of water level meter calibration per year.
Op11 –	On-line water quality monitoring equipment calibration (-/year)
	Rate of on-line water quality monitoring instrument calibration per year.

Inspection and maintenance procedures should include electrical and signal transmission equipment:

Table 17. Electrical and signal transmission equipment inspection PI (II #224).

Op12 –	Emergency power system inspection (-/year)
	Rate of power systems inspection per year assessed in terms of nominal power.
Op13 –	Signal transmission equipment inspection (-/year)
	Rate of signal transmission units inspection per year.
Op14 –	Electrical switchgear equipment inspection (-/year)
	Rate of electrical switchgear units inspection per year.

Mobility is an important aspect for the operation and maintenance of water services. It can be assessed in terms of vehicle availability:

Table 18. Vehicle availability PI (II #225).

Op15 –	Vehicle availability (No./100 km)
	Number of vehicles available daily, on a permanent basis, on average, for field works in operation and maintenance activities per 100 km of mains.

Long-term performance and sustainability assessment are closely related to rehabilitation practices. The following indicators relate to mains, valves, service connections and pumping stations rehabilitation. See the definitions chapter to clarify the concepts related to rehabilitation, renovation, replacement and refurbishment that shall be adopted when applying the IWA-PI systems.

Table 19. Rehabilitation PI (II #226).

Mains, valves and service connection rehabilitation	
Op16 –	Mains rehabilitation (%/year)
	Percentage of mains length rehabilitated per year.
	Op17 – Mains renovation (%/year)
	Percentage of mains length renovated per year.
	Op18 – Mains replacement (%/year)
	Percentage of mains length replaced per year.

Table 19. Rehabilitation PI (II #226) (contd.).

	Op19 – Replaced valves (%/year)
	Percentage of mains valves replaced per year.
Op20 –	Service connection rehabilitation (%/year)
	Percentage of service connections replaced or renovated per year.

Table 19.a. Rehabilitation PI (II #166) (contd.).

Pumps rehabilitation	
	Op21 – Pump refurbishment (%/year)
	Percentage of pumps that were subject to overhaul per year, assessed in terms of nominal power.
	Op22 – Pump replacement (%/year)
	Percentage of pumps replaced per year, assessed in terms of nominal power.

Water losses control is one of the major management concerns. It is very important to use appropriate performance indicators to appropriately monitor the situation from the operational viewpoint. The operational water losses indicators recommended by IWA are:

Table 20. Operational water losses PI (II #229).

Op23 –	Water losses per connection (m^3/connection/year)
	Total (apparent and real) losses, expressed in terms of annual volume lost per service connection. This indicator is adequate for urban distribution systems.
Op24 –	Water losses per mains length (m^3/km/day)
	Total (apparent and real) losses, expressed in terms of annual volume lost per mains length. This indicator is adequate for bulk supply and low service connection density distribution systems.
	Op25 – Apparent losses (%)
	Percentage of the water provided to the system (system input volume minus exported water) that corresponds to apparent losses. This indicator is adequate for urban distribution systems.

Table 20. Operational water losses PI (II #229) (contd.).

	Op26 – Apparent losses per system input volume (%)
	Percentage of the water entering the system (exported water inclusive) that corresponds to apparent losses. This indicator is adequate for bulk supply and low service connection density distribution systems.
	Op27 – Real losses per connection (l/connection/day when system is pressurised)
	Real losses, expressed in terms of the average daily volume lost per connection. This indicator is adequate for urban distribution systems.
	Op28 – Real losses per mains length (l/km/day when system is pressurised)
	Real losses, expressed in terms of the average daily volume lost per mains length. This indicator is adequate for bulk supply and low service connection density distribution systems.
Op29 –	Infrastructure leakage index (-)
	Ratio between the actual real losses and an estimate of the minimum real losses that could be technically achieved for the system operating pressure, average service connection length and service connection density.

The correct assessment of these indicators requires a proper interpretation of the terminology and the implementation of the recommended water balance procedure. See II-1.1 for detailed information.

Op23 is less relevant than the indicators directed specifically to the apparent and to the real losses performance. However, sometimes it is the only one of this group that the *utilities* are able to assess. In fact, the interventions required to decrease each of the two types of losses are very different, and it is well worth the effort to try and get the best estimate of the portion of the total losses corresponding to each type.

In the First Edition of this manual, the apparent losses were expressed in terms of volume lost per connection (for the distribution systems) or per mains length (for the bulk supply systems). The field test and the experience of the

IWA water losses taskforce demonstrated that this was not a good option, and the use of percentages is now recommended.

The real losses indicator Op27 is a much better operational performance measure than the traditional percentage indicator. Practice demonstrates that the density of service connection is an important factor affecting the leakage volume. Removing its influence allows for creating a better indicator for comparisons and result interpretation. Op28 is the corresponding indicator for the cases where the density of service connections is very low, and therefore the mains length becomes a dominating explanatory factor.

A similar explanation applies to the infrastructure leakage index (ILI – Op29), which aims to remove other factors not related to the physical condition of the network that significantly affect the leakage volume. These are the cases of operating pressure and of average length of the service pipes (from the property boundary – delivery point – to the measurement point). It is assessed on the basis of an empirical expression proposed by the IWA water losses task force which compares the actual real losses with an estimate of the real losses that a well-constructed and maintained system with the same service connection density and average length and the same average operating pressure would have. This indicator generated a lot of discussion during the field test. It has great support, but also received a lot of criticism. In general, it seems to be supported by water losses consultants. Some *utilities* also expressed the view that it is useful for comparison of different district metering areas (DMA), allowing for prioritising interventions. The criticisms are of a various nature. The first one, supported by the authors, is that it is the only indicator of the whole IWA PI systems that contains a judgement in itself and is based on an empirical expression (and for these reasons does not fit all the PI requirements (see I-2.2). The other shortcomings relate to its meaning and confidence level when the variability of the operating pressure and/or the service connection length in the system is high (e.g., hilly regions, systems with significant daily pressure fluctuations, systems with apartment blocks and individual apartment meters). The important thing is that users are well aware of both its potential and shortcomings when selecting and applying it.

The next group of indicators are measures of failures. They are particularly relevant for asset management.

Table 21. Failure PI (II #234).

Op30 –	Pump failures (days/pump/year)
	Average number of days per year system pumps are out of order.
Op31 –	Mains failures (No./100 km/year)
	Average number of mains failures per 100 km of mains and per year.
Op32 –	Service connection failures (No./1000 connections/year)
	Average number of service connection failures, expressed per 1000 connections and per year.
Op33 –	Hydrant failures (No./1000 hydrants/year)
	Average number of hydrant failures, expressed per 1000 hydrants and per year.
Op34 –	Power failures (hours/pumping station/year)
	Average number of hours per year pumping stations are out of service due to power supply interruptions.
Op35 –	Water-point failures (No./water-point/year)
	Average number of failures per water-point and per year.

The frequency of asset failures affects directly the quality of the service provided and the maintenance costs. Conversely – and this is a relevant aspect from the operational point of view – they are an indirect measure of the assets physical condition.

Op34 and Op35 are indicators proposed by developing regions, where it is usual to find frequent powers supply problems (electrical supply interruptions or lack of fuel) and inoperative water supply points. These are therefore very important performance indicators in these regions.

The next group of indicators deals with water metering performance:

Table 22. Water metering PI (II #237).

Op36 –	Customer reading efficiency (-)
	Ratio between the effective meter readings and the total number of readings if all the customers meters were read according to the pre-established frequency.

Table 22.a. Water metering PI (II #174) (contd.).

Op37 –	Residential customer reading efficiency (-)
	Ratio between the effective residential meter readings and the total number of readings if all the residential customers meters were read according to the pre-established frequency.
Op38 –	Operational meters (%)
	Percentage of the customer meters that are installed that are not out-of-service.
Op39 –	Unmetered water (%)
	Percentage of system input value that is not accounted for as metered consumption.

There are factors that lead to lack of meter reading efficiency. In some countries, for instance, there were times when the meters were installed in households, with difficult access by the undertaker. Indicators Op36 and Op37 aim to track this type of situation. Op38 tackles a very important problem in some developing regions, where it is frequent to find a high portion of the installed meters out of service. It does not account for the working inaccurate meters, but it is a simple to assess and effective indicator, recommended and successfully tested in the field test.

Op39 was also recommended during the field test. Although it is not a water losses indicator, it provides relevant information in the context of water losses.

The next group of indicators deal with the performance of monitoring the quality of the water delivered to the customers.

Table 23. Water quality monitoring PI (II #239).

Op40 –	Tests carried out (%)
	Percentage of treated water tests required by applicable standards or legislation that are carried out.
	Op41 – Aesthetic tests carried out (%)
	Percentage of aesthetic tests of treated water required by applicable standards or legislation that are carried out.

Table 23. Water quality monitoring PI (II #239) (contd.).

	Op42 – Microbiological tests carried out (%)
	Percentage of microbiological tests of treated water required by applicable standards or legislation that are carried out.
	Op43 – Physical-chemical tests carried out (%)
	Percentage of aesthetic tests of treated water required by applicable standards or legislation that are carried out.
	Op44 – Radioactivity tests carried out (%)
	Percentage of radioactivity tests of treated water required by applicable standards or legislation that are carried out.

These indicators are organised in function of the nature of the water parameters. This classification may not coincide with the classification of the applicable legislation. However, a coherent and general classification that could be adopted by any country had to be adopted, not necessarily the easiest one to apply for particular instances.

These indicators are directed to the quantity of tests. Their results, i.e., the compliance aspect, is considered in the quality of service section.

At first glance, it may seem that there were no changes to these indicators in comparison with the First Edition of the manual, which is not the case. With the underlying assumption that a small number of water quality indicators would be considered[4] there was a need to use aggregated information. The indicators of the First Edition were assessed as the ratio between the total number of tests performed within a category and the total number of tests required for that category. The consequence was that if, for a given parameter, the *utility* performed more tests than required, this would "compensate" for the lack of monitoring in other parameters of that category and the indicator did not reflect that. Results of more than 100% could easily be achieved for the indicators even though a good number of parameters were under-monitored. The indicators defined in such a way proved to be misleading and meaningless. Currently, the number of tests performed for any parameter to be considered for the calculation is limited to a maximum equal to the respective number of

[4] It was out of the question to consider one PI for parameter, even if a limited number of parameters were chosen; the users are free and encouraged to split these aggregate indicators according to their needs.

tests required. If the users are interested in tracking whether or not more tests than required are performed, new and parameter-specific PI should be added.

Quality of service indicators

When the core business is to provide a water supply service, the assessment of the quality of this service is an obvious management need. Quality of service indicators recommended aims are to know the extent of the service coverage within the area of influence of the undertaker and the performance of the service in terms of quantity and quality of water provided.

The service coverage indicators offer some alternatives for assessment, dependent on local characteristics. All of them provide a service coverage percentage.

Table 24. Service coverage PI (II #241).

QS1 –	Households and businesses supply coverage (%)
	Percentage of existing households and businesses that are connected to the public network.
QS2 –	Buildings supply coverage (%)
	Percentage of existing buildings that are connected to the public network.
QS3 –	Population coverage (%)
	Percentage of the resident population that is served by the *utility*.
	QS4 – Population coverage with service connections (%)
	Percentage of the resident population that is served through service connections.
	QS5 – Population coverage with public taps or standpipes (%)
	Percentage of the resident population that is served through public taps or standpipes.

The most popular of these indicators is QS3. However, published statistics using it do not provide a clear associated definition. Conversely, *utilities* do not have accurate information on how many people they serve. They probably have an accurate record of how many customers of each type, but they typically transform these figures into inhabitants adopting an average household dimension. This procedure has in general poor accuracy. It is particularly problematic in regions with significant seasonal fluctuations of population. It is therefore recommended that a more objective and reliable alternative be used. In countries where the practice is to have one customer per household (or business connection), QS1 is the recommended option. It expresses the percentage of existing households and businesses that have access to the service, regardless of being for permanent or temporary use. In countries or regions where there is a single customer per building, QS2 is the recommended alternative. It has the disadvantage of giving the same weight to apartment blocks as to detached houses, but nevertheless it is in general much more reliable than QS3. This latter option is left as the recommended alternative for situations where there are other forms of service beyond the connection to the public network (i.e., public taps and standpipes). This indicator should be assessed based on the resident population. This is because practice shows that statistics on seasonal population – either total or served – are rather inaccurate.

A global coverage of a country or region may not be deduced directly from the individual *utilities* coverage values because situations exist, particularly in developing regions, where there are households located out of the area of influence of any existing *utility*.

In developing regions, a significant part of the population is served through public taps and standpipes. The following indicators are measures of the performance of this type of service.

Table 25. Public taps and standpipes PI (II #244).

QS6 –	Operational water points (%)
	Percentage of existing water points that are not out of service.
QS7 –	Average distance from waterpoints to households (m)
	Average distance between water points and the respective far-most household served by it.
QS8 –	Per capita water consumed in public taps and standpipes (l/person/day)
	Ratio between the volume of water supplied by public taps and standpipes and the population they supply.
QS9 –	Population per public tap or standpipe (persons/tap)
	Average number of persons served by public tap or standpipe.

From the quantitative point of view, a good quality service has adequate pressure, is available 24 hours per day, without interruptions or restrictions. These are the aspects covered by the next group of PI.

Table 26. Pressure and continuity of supply PI (II #245).

QS10 –	Pressure of supply adequacy (%)
	Percentage of delivery points (one per service connection) that receive and are likely to receive adequate pressure.
QS11 –	Bulk supply adequacy (%)
	Percentage of delivery points that are supplied at any time according to the target flow, volume and/or pressure.
QS12 –	Continuity of supply (%)
	Percentage of hours when the (intermittent supply) system is pressurised.
QS13 –	Water interruptions (%)
	Average percentage of hours each person served is subject to water interruptions.
QS14 –	Interruptions per connection (No./1000 connections/year)
	Average number of interruptions per service connection per year.
QS15 –	Bulk supply interruptions (No./ delivery point /year)
	Average number of interruptions per delivery point per year.
QS16 –	Population experiencing restrictions to water service (%)
	Average percentage of hours each served person experienced restrictions to water service.
QS17 –	Days with restrictions to water service (%)
	Percentage of days with restrictions to water service.

QS12 refers to intermittent supply systems and aims to assess the portion of the day or of the week the population has access to piped water, whereas the interruption indicators refer to irregular interruptions due to system failures. This seemed to be a suitable set of indicators to deal with reliability of supply. However, in some cases it may be unclear if we are facing a continuity problem or an interruption. This is what happens, for instance, in permanent supply systems which may fail to deliver water continuously during some odd periods of the year, due to lack of water or lack of capacity of some components of the system. In these situations, the definitions provided may need to be completed, in order to prevent misinterpretation[5].

QS13 and QS16 provide more information than QS14 and QS17, their respective alternatives. However, their assessment is more demanding in terms of the data required, and not every *utilities*' manager might have these data. The field test demonstrated that QS14 and QS17 could be assessed by most participants.

Another fundamental aspect of the quality of the service is the quality of the water. This issue is considered in the next group of indicators, complementary to the operational indicators Op40 to Op44, which deal with water quality monitoring.

Table 27. Quality of supplied water PI (II #249).

QS18 –	Quality of supplied water (%)
	Percentage of the total number of treated water tests performed that comply with the applicable standards or legislation.
	QS19 – Aesthetic tests compliance (%)
	Percentage of the total number of aesthetic tests of treated water performed that comply with the applicable standards or legislation.
	QS20 – Microbiological tests compliance (%)
	Percentage of the total number of microbiological tests of treated water performed that comply with the applicable standards or legislation.
	QS21 – Physical-chemical tests compliance (%)
	Percentage of the total number of physical-chemical tests of treated water performed that comply with the applicable standards or legislation.
	QS22 – Radioactivity tests compliance (%)
	Percentage of the total number of radioactivity tests of treated water performed that comply with the applicable standards or legislation.

5 For instance in Portugal, where there are not intermittent supply systems, the Regulator decided to merge the assessment of continuity and interruptions into a single indicator (case study described in 5.3).

Contrary to what happens in the PI Op40 to Op44, in this case the number of tests considered is not limited by the applicable legislation.

Service connection and meter installation and repair are also relevant aspects of the quality of the service provided to customers:

Table 28. Service connection and meter installation and repair PI (II #251).

QS23 –	New connection efficiency (days)
	Average time spent from customer requests until the availability of the water service, for existing service connection assets.
QS24 –	Time to install a customer meter (days)
	Average time spent from customer requests for the installation of newly fitted water meters until the availability of the service.
QS25 –	Connection repair time (days)
	Average time spent repairing service connections.

QS24 is an indicator recommended during the field test. It is relevant in situations where there is a policy of implementation of customer meters. In general, the installation of newly fitted meters requires some construction work.

All the previous PI are measures of the quality of the service provided, but they do not reflect customers views. The analysis of customer complaints is also very important:

Table 29. Customer complaints PI (II #252).

QS26 –	Service complaints per connection (No. complaints/1000 connections/year)
	Average number of complaints of quality of service per 1000 service connections and per year. This indicator is adequate for distribution systems.
QS27 –	Service complaints per customer (No. complaints/customer/year)
	Average number of complaints of quality of service per 1000 service connections and per year. This indicator is adequate for bulk supply and low service connection density distribution systems.

Table 29. Customer complaints PI (II #252) (contd.).

	QS28 – Pressure complaints (%)
	Percentage of service complaints regarding pressure problems.
	QS29 – Continuity complaints (%)
	Percentage of service complaints regarding continuity problems.
	QS30 – Water quality complaints (%)
	Percentage of service complaints regarding water quality problems.
	QS31 – Interruption complaints (%)
	Percentage of service complaints regarding interruption problems.
QS32 –	Billing complaints and queries (No./customer/year)
	Average number of billing complaints and queries per customer and per year.
QS33 –	Other complaints and queries (No./customer/year)
	Average number of other complaints and queries per customer and per year.
QS34 –	Response to written complaints (%)
	Percentage of written complaints responded to within the target time.

Complaints PI are valid and relevant, but need to be interpreted with care. In fact, they are not direct performance measures. Whereas in most cases the evolution of a PI with time allows direct understanding of whether the performance, for that specific aspect, is decreasing or increasing, this may not apply to complaint PI, if customer service changes occur at the same time. Customers do not complain if they do not have easy access to their service provider or if they know their complaint is not taken into account. Therefore, a natural evolution of complaints in services with a growing quality of service is an increase in complaints when customers realise they can be heard, followed by a period of decrease in complaints when the service problems are sorted out followed by a phase of stabilisation.

If the analysis of an individual PI, isolated and out of context, is dangerous and potentially misleading, for this group of PI this is particularly the case. On the one hand, explanatory factors about the reasons affecting the number of complaints are needed; on the other hand, these indicators need to be observed with other related PI, such as pressure adequacy, continuity, interruptions, water quality or failures rates.

Economic and financial indicators

The last group of PI deals with the economic and financial performance of a utility. It contains firstly, revenues, costs, investment, water charges, and more descriptive economic and financial indicators and secondly efficiency, leverage, profitability and water loss indicators.

The assessment of these indicators requires compliance with the financial definitions of section II-1.3, set up according to international accounting rules. The recent adoption of the International Financial Report Standards (IFRS) may have potential impacts on both the quality of the input variables and on the values of the PI assessed. Therefore, any comparison among peers need extra care. Prior harmonization may be required, particularly if some them have not adopted yet the IFRS principles.

The implementation in a given country may require some degree of adaptation, but the *utilities* who really tried to assess these indicators during the field test ended up being successful without great difficulty.

One of the major difficulties had to do with the proper translation of the terms into local languages, fundamental in this case for a clear understanding of the concepts. The availability of the IWA PI system in English, French, German, Chinese, Persian, and Portuguese, with partial translations in other languages (e.g., Spanish, Japanese, Czech language) has helped the future implementation in a good number of countries. Both the field-test and the preparation of the translations offered the opportunity for finding out the terms and definitions not sufficiently clear in the First Edition of the manual and hence solved the problem.

The following tables present the indicators recommended, starting with the classical revenue indicators:

Table 30. Revenue PI (II #256).

Fi1 –	Unit revenue (EUR/m^3)
	Revenue per cubic metre of authorised consumption.
	Fi2 - Sales revenues (%)
	Percentage of revenues coming from sales.
	Fi3 - Other revenues (%)
	Percentage of revenues not coming from sales.

The particular strength of these indicators is the detailed specification of the input variables. The total revenues should not include the capitalised costs of self-constructed assets; the authorised water consumption includes not only the billed water but also other authorised consumption that is not billed (e.g., firefighting, etc.).

The basic cost indicators are also the traditional ones:

Table 31. Cost PI (II #257).

Fi4 –	Unit total costs (EUR/m³)
	Total costs (running & capital) per cubic metre of authorised consumption.
	Fi5 – Unit running costs (EUR/m³)
	Running costs per cubic metre of authorised consumption.
	Fi6 – Unit capital costs (EUR/m³)
	Capital costs per cubic metre of authorised consumption.

Running costs should be expressed at their net value (obtained by negative apportionment of related capitalised costs of self-constructed assets). In any case, the interpretation of the unit costs achieved may require in-depth analysis of the composition of the running costs per main type of costs:

Table 32. Composition of running costs per type of costs PI (II #258).

	Fi7 – Internal manpower costs (%)
	Percentage of the running costs corresponding to internal manpower.
	Fi8 – External services costs (%)
	Percentage of the running costs corresponding to external services.
	Fi9 – Imported (raw and treated) water costs (%)
	Percentage of the running costs corresponding to imported water.
	Fi10 – Electrical energy costs (%)
	Percentage of the running costs corresponding to electrical energy.
	Fi11 – Other costs (%)
	Percentage of the running costs corresponding to other costs not considered in Fi7 to Fi10.

In an earlier version of the PI system, one category was "Energy costs", as it seemed to be the most logical solution. However – and this is just an example of balances and decisions made during the process of building the PI system – practice demonstrated that the assessment of this indicator would be too time consuming to assess. The reason is that accounting rules separate electricity costs from fuel, which is included in "purchased merchandises". The decision was to find relevant categories which are compatible with accounting rules.

The above indicators allow the identification of "big figures" and the comparison with values from peer organisations (taking the context into account), and hence to try and identify where potential for savings are.

Another complementary dimension of the same type of analysis is the observation of the running costs per main function, in the categories as adopted for the Personnel PI:

Table 33. Composition of running costs per main function of the *water utility* PI (II #260).

	Fi12 – General management functions costs (%)
	Percentage of the running costs corresponding to running costs of general management functions.
	Fi13 – Human resources management functions costs (%)
	Percentage of the running costs corresponding to running costs of general management functions.
	Fi14 – Financial and commercial functions costs (%)
	Percentage of the running costs corresponding to running costs of financial and commercial functions.
	Fi15 – Customer service functions costs (%)
	Percentage of the running costs corresponding to running costs of customer service functions.
	Fi16 – Technical services functions costs (%)
	Percentage of the running costs corresponding to running costs of technical services functions.

There were more *utilities* reporting difficulties with indicators Fi12 to Fi16 than with the previous ones, because very few had accounting systems that allowed them to obtain the input variables. However, the field-test participants confirmed the relevance of these indicators and supported their inclusion in the system, as well as the following one, which goes into more detail with regard to the technical functions.

Table 34. Composition of running costs per technical function activity PI (II #261).

	Fi17 – Water resources and catchment management costs (%)
	Percentage of the running costs corresponding to running costs of the water resources and catchment management.
	Fi18 – Abstraction and treatment costs (%)
	Percentage of the running costs corresponding to running costs of abstraction and treatment.
	Fi19 – Transmission, storage and distribution costs (%)
	Percentage of the running costs corresponding to running costs of transmission, storage and distribution.
	Fi20 – Water quality monitoring costs (%)
	Percentage of the running costs corresponding to running costs of water quality monitoring.
	Fi21 – Meter management costs (%)
	Percentage of the running costs corresponding to running costs of meter management.
	Fi22 – Support services costs (%)
	Percentage of the running costs corresponding to running costs of support services.

The capital costs decomposition follows a different – much simpler structure:

Table 35. Composition of capital costs PI (II #263).

	Fi23 – Depreciation costs (%)
	Percentage of the capital costs corresponding to depreciation costs.
	Fi24 – Net interest costs (%)
	Percentage of the capital costs corresponding to net interest costs (i.e., difference between interest expenses costs and interest income).

The important thing is to differentiate the investment for new assets and reinforcement (e.g., expansion) of existing assets from the investment for asset replacement and renovation. The analysis of these indicators is most relevant in the scope of asset management.

Table 36. Investment PI (II #264).

Fi25 –	Unit investment (EUR/m^3)
	Cost of investments (expenditures for plant and equipment) per cubic metre of authorised consumption.
	Fi26 – Investments for new assets and reinforcement of existing assets (%)
	Percentage of the cost of investments corresponding to investments in new assets and reinforcement of existing assets.
	Fi27 – Investments for asset replacement and renovation (%)
	Percentage of the cost of investments corresponding to investments for asset replacement and renovation.

When concessions exist, the International Financial Reporting Standards, IFRS, account for the assets constructed or upgraded according to the concession arrangement among the Intangible Assets and depreciate them, on a systematic basis, over the concession period (as detailed in section II-1.3). Given that in many cases water utilities traditionally accounted, and some still do, the assets constructed or upgraded according the concession arrangement as Water service Assets and depreciated them according to different concepts of useful lives, careful attention is recommended when comparing utilities.

The final set of this first group of more descriptive economic and financial indicators regards the water charges, split into charges for bulk supply and for direct consumers. As the tariff structure depends on case to case, the solution adopted, of universal application, is to assess aggregated indicators that include all the charges to all customers in the numerator and the sum of all the volumes provided to the users (billed or unbilled) in the denominator.

Table 37. Average water charges PI (II #266).

Fi28 –	Average water charges for direct consumption (EUR/m^3)
	Water sales revenue from residential, industrial and other customers (exported water excluded; public water taxes excluded) per cubic metre of authorised consumption, exported water excluded.
Fi29 –	Average water charges for exported water (EUR/m^3)
	Water sales revenue from exported water (excluding public water taxes) per cubic metre of exported water.

The next set of indicators allows the analysis of the efficiency of the *utility.*

Table 38. Efficiency PI (II #267).

Fi30 –	Total cost coverage ratio (-)
	Ratio between the total revenues and the total costs.
Fi31 –	Operating cost coverage ratio (-)
	Ratio between the total revenues and the running costs.
Fi32 –	Delay in accounts receivable (days equivalent)
	Ratio between the accounts receivable from drinking water and the sales revenues.
Fi33 –	Investment ratio (-)
	Ratio between the investments subject to depreciation and the depreciation costs.
Fi34 –	Contribution of internal sources to investment = CTI (%)
	Percentage of total investments that are financed by the cash flow.
Fi35 –	Average age of water service assets (%)
	Percentage of the historical value of water service assets that correspond to depreciated historical value of water service assets.
Fi36 –	Average depreciation ratio (-)
	Ratio between the depreciation costs and historical value of water service assets.
Fi37 –	Late payments ratio (-)
	1 minus the ratio between the annual debt from customers and the amount billed.
Fi38 –	Inventory value (-)
	Ratio between the value of overall inventory at the end of the fiscal year and the operating revenues.

Some of the above indicators correspond to classical economic efficiency indicators, while others do not. The result is the outcome of very intensive team work based on the indicators of the First Edition. No fundamental changes were made, but the minor changes have made a difference in terms of clarity and acceptability.

The following set deals with leverage:

Table 39. Leverage PI (II #270).

Fi39 –	Debt service coverage ratio = DSC (%)
	Percentage of the financial debt service that correspond to cash-flow.
Fi40 –	Debt equity ratio (-)
	Ratio between total debt and the shareholders' equity.

Please note that in some countries the debt equity ratio is assessed as the reverse of Fi40 (i.e., ratio between the shareholders' equity and the total debt). Care must be taken when comparing values across different countries.

The assessment and interpretation of these indicators proved to be sometimes problematic in situations when there are external subsidies for the investment (e.g., international donations in developing regions; cohesion funds within the European Union). A well informed analysis is required, and even some adaptation of the indicators' processing rules may be needed.

Liquidity can be assessed using the current ratio:

Table 40. Liquidity PI (II #270).

Fi41 –	Current ratio (-)
	Ratio between current assets and current liabilities.

A set of four indicators is recommended to asses profitability:

Table 41. Profitability PI (II #271).

Fi42 –	Return on net fixed assets (%)
	Percentage of the historical value of the *water utility* assets, deduced from the depreciated historical value of *water utility* assets, that correspond to operating income
Fi43 –	Return on equity (%)
	Percentage of the shareholders' equity that corresponds to net income (after interest payment and taxes).
Fi44 –	Return on capital employed (%)
	Percentage of the total assets that correspond to operating income (EBIT) (deduced from related taxes).
Fi45 –	Asset turnover ratio (-)
	Ratio between sales revenues and total assets.

The final group of indicators come back to the issue of water losses, this time from the economic point of view.

Table 42. Economic water losses PI (II #272).

Fi46 –	Non-revenue water by volume (%)
	Percentage of the system input volume that corresponds to non-revenue water.
Fi47 –	Non-revenue water by cost (%)
	Percentage of the system input volume that corresponds to the valuation of non-revenue water components.

Fi46 is perhaps the most popular and easy way to assess water losses PI. Fi47 is an innovative indicator recommended by the IWA Water Losses Task Force which was considered rather interesting by the participants in the field-test. However, it is not easy to assess, particularly if the area under analysis does not cover the whole area of influence of the *utility*. It is difficult to accurately allocate current costs to sub-systems or district metering areas.

5.5 Variables

As explained in I-2.1, variables are the data elements that are necessary to assess the PI, in compliance with the requirements listed in I-2.2.

Variables are organised in groups and subgroups, as referred in I-5.2. Variables are listed and defined in detail in II-3.

It is very important to note that just a selection of the variables is required for any specific implementation of a PI system. Only after the preselection of the relevant PI should the users identify the subset of variables that are needed. During the field test, some participants did the reverse: they filled in as many variables as they were able, and then realised which PI they could use. This procedure is acceptable for the sake of testing the system, but should not be adopted as normal practice. The recommended implementation procedure is described in chapter I-6.

There are variables which are used for one single specific indicator, and others which are used for many indicators. Variable forms of chapter II-3 list the indicators they are required for.

Almost all variables are used directly in, at least, one PI, as primary variables. There are some that are also used to assess other variables, playing the role of secondary variables.

As for PI, all the variables refer to the core business of a water supply service. Multi-service utilities should assess the part of resources and revenues related to the water supply component. The same applies to *utilities* with complementary businesses, such as for instance, selling services related to water quality sampling and testing or repair of private property plumbing.

Regardless of the software that the users may adopt for data input and PI assessment, it should be prepared to handle the three fundamental pieces of information for each variable that the IWA PI system requires: the value of the variable, the accuracy band and the reliability of the data source (see I-6 and II-5):

Table 43. Example of variable input table.

Variable	Value	Accuracy band	Reliability of the data source
C8 – Mains length (km)	1398	0–5%	***
C21 – Mains valves (No.)	10027	0–5%	***
C24 – Service connections (No.)	93000	5–20%	**

This is a characteristic of the IWA PI systems that other existing PI systems do not have. Practice using the system in different contexts has demonstrated how important this is for the awareness of the confidence of the information handled and of the need to invest in data quality control. *utilities* participating in the field test reported that it is worth the effort, provided that there are not too many categories to choose from. This is the reason why the number of categories in the First Edition was reduced for the Second Edition.

One of the greatest challenges of the IWA PI projects was to find consensus on clear definitions for the variables. It is very easy for different people to have different interpretations of the same text.

For some variables this is an easily achievable objective, but not for others. To cope with these more complex cases, the following definition scheme was established:

- A methodology to assess the water balance components is recommended, in accordance with the IWA Water Losses Task Force (II-1.1)
- The functions and sub-functions of a water supply service are defined in the way that they should be adopted for the standardised assessment of the personnel and the economic and financial variables and PI. In general, the organisation organogram does not fully coincide with the aggregation of sub-functions into functions. This is neither the aim nor a problem, as the field test demonstrated. Some adjustments have been made with regard to the First Edition of the manual, according to experience.(II-1.2).
- A detailed definition of economic and financial concepts, set up according to the international accounting rules. The field test proved to be most useful for the improvement of these definitions (II-1.3).

- A glossary of complementary definitions, shared by various variables and PI (II-1.4).
- A list of variable-specific definitions and comments, contained in the variable forms (II-3).

5.6 Explanatory factors

Types of explanatory factors

We learned in I-2.1 that an explanatory factor is any element of the system of performance indicators that can be used to explain PI values, at the analysis stage. This includes PI, variables, context information and other data elements not playing an active role before the analysis stage.

There are explanatory factors that depend on short and medium term management options. Their identification and analysis is crucial for the identification and prioritization of improvement measures. This type of explanatory factor is also known as a "driving factor". There are other explanatory factors that are contextual, and cannot be changed by the *utility*, at least in the short or medium term. In some cases, the differences in context are so relevant that they prevent any valid PI comparisons.

It should be noted that explanatory factors are PI dependent, i.e., one explanatory factor that is relevant to one PI or a set of PI may be irrelevant to others. The main consequence of this fact is that two different *utilities* may be comparable for some indicators and not for others.

In the establishment of the IWA PI system, one of the underlying objectives was to minimise the dependence on the context of each PI. For instance, indicators potentially relevant for use within a *utility* but which are too context dependent (e.g., per capita consumption or energy consumption per cubic meter) were not included in the system. However, full context independence is not a feasible target for any PI system.

The following sections deal with several types of explanatory factors:

- Context information
- Performance indicators and variables
- Other explanatory factors

Context information

Context information is especially useful in benchmarking initiatives. In a preliminary stage, it supports the establishment of clusters of comparable *utilities*. In the analysis stage, it should be taken into account in order to identify whether differences in performance values correspond to differences in management procedures or in context.

Internally, context information is important for understanding variations of performance over time.

Context information may also be used to test given hypotheses for the explanation of performance. Questions such as "How does the type of ownership or the type of operations affect performance?", "What is the influence of system and *utility* size on the performance?" or "Do multi-service *utilities* tend to be more efficient than single service *utilities*?" require a lot of care when being tested. There are not two *utilities* that have identical context information except for the one aspect being tested, and therefore independence and sound approaches are fundamental.

In general, context information is relatively stable. Demography, climate, economy, system characteristics, etc. do not change overnight. This is a general characteristic of context information. However, there are also some types of context information that may have a much more dynamic nature. This is the case for special events which may affect temporarily the performance of a system. Typical examples are weather (i.e., exceptional floods or droughts), crisis situations or exceptional concentrations of people (i.e., the funeral of Pope John Paul II, in Rome).

As the field test of the IWA PI system was not a benchmarking initiative, there were practical constraints to a thorough and systematic testing of the context information recommended. However, suggestions for improvement were taken into account. One important point is to what level of detail and quantity context information should be specified. The IWA PI system contains detailed and quantified specifications. The German case study reported in I-8.5 is a successful example of the use of quantified explanatory factors. On-going experiences adopt different solutions. For instance the Portuguese regulator, ERSAR, adopted qualitative classes of "context factors" in its PI system (example 2,2, I-8.2). This is still a matter for discussion and further development, objective of the European COST Action C18 – *Performance assessment of urban infrastructure services: the case of water supply, wastewater and solid waste* (www.costc18.org).

The full list of context information of the IWA PI system is presented and defined in II-4. It is structured into six main groups, as shown in Table 44.

Table 44. Types of context information of the IWA PI system.

***Utility* profile** (page II-368)	
Service information (page II-370)	
System assets (page II-372)	
	Water resources
	Raw water reservoirs
	Treatment plants
	Treated water reservoirs
	Pumping stations
	Transmission and distribution network
	Service connections
	Private pumping systems and tanks
Consumption and peak factors (page II-379)	
	Average system input
	Consumption per type of customer
	Peak factors of supplied and exported water
Demography and economics (page II-381)	
Environment (page II-383)	
	Annual rainfall
	Air temperature
	Topography
	Raw water quality – Source types

Some of the above listed types of context information are external and totally independent from the *utility*'s management decisions, such as environmental information. Others, such as demography and economics are almost independent; although the provision of the water supply service is very important for the economy of a region, it can also be accepted that the *utility* by itself has a limited capacity to change significantly demography and economics.

Some of the context information is a combination of external and management factors. This is for instance the case for consumption and peak factors, taking into account that the *utility* has significant potential to manage demand. This is also the case for service context information.

System assets context information depends on long term policies of the *utility*. The situation assessed today is affected by the past policies of the *utility*. Changes in policies do not have immediate effects either. However, measures for improvements should include this long term perspective.

The *utility* profile depends on political decisions. The asset owner may or may not have the capacity to modify it, on a case to case basis.

Performance indicators and variables as explanatory factors

The performance analysis of an *utility* as a whole, or of a given process, needs to be done based on a set of inter-related PI, some being explanatory factors of the others. For instance, it is likely that *utilities* that invest in system preventative maintenance, inspection and rehabilitation will have lower failure rates, service interruptions, and water losses. There is also an inter-relationship between one PI and its parent or child PI (e.g., if the PI for the total number of employees is very high, it would be rational to find out whether this is due to any specific function).

There are also cases where the interpretation of the performance values find the explanation for certain results in data elements which are also variables (i.e., used to assess PI). For instance, the comparison between a big and a small utility can be held, but the interpretation should take these differences into account, as they may explain differences in the results. Similarly, the adoption of IAS principles highly influences key financial variables and reports; comparing PIs of different utilities, where just some of them are adopting IAS compliant accounting systems, may be misleading due to the lack of common concepts and definitions. Table 45 presents examples of variables which reflect the size of the *utility* and of the system, which are sometimes relevant as explanatory factors.

Table 45. Examples of variables that may be relevant as explanatory factor.

Size of the *utility*	
G4 –	Annual costs (EUR/year)
G1 –	Total revenues (EUR/year)
G32 –	Average annual investment (EUR/year)
	(a moving average of the last 3–5 years should be used)
B1 –	Total personnel (No.)
Size of the system	
C8 –	Mains length (km)
E6 –	Customer flow meters (No.)
C10 –	System flow meters (No.)
C22 –	Valves (No.)
C23 –	Hydrants (No.)
F8 –	Public taps and standpipes (No.)

Other explanatory factors

There may be other factors affecting the PI values in a significant way that are neither part of the context, because they may be changed by managers in the short term, nor variables nor PI. Some may explain a better or poorer performance, others may simply explain differences in values, not meaning that the real performance is different.

The IWA PI system does not specify a list of other explanatory factors, because they are case dependent. However, this section provides some examples.

One of the most relevant and fundamental for the comparison of personnel PI, is the level of outsourcing. A *utility* which outsources a significant part of the activity will have naturally a lower rate of personnel for the same efficiency. Table 46 contains some recommendations on how to express outsourcing.

Table 46. Other explanatory factors: level of outsourcing.

	General management (%)
	Estimated cost percentage of the general management function (dealing with water supply) that is outsourced.
	Human resources management (%)
	Estimated cost percentage of the human resources management function (dealing with water supply) that is outsourced.
	Financial and commercial (%)
	Estimated cost percentage of the financial and commercial function (dealing with water supply) that is outsourced.
	Customer service (%)
	Estimated cost percentage of the customer service function (dealing with water supply) that is outsourced.
	Planning and design (%)
	Estimated cost percentage of the planning and design function (dealing with water supply) that is outsourced (e.g., use of consultants).
	Construction (%)
	Estimated cost percentage of the construction function (dealing with water supply) that is outsourced (proportion by value of new works and asset refurbishment undertaken by external contractors, including new service connections and extensions of network).
	Operations and maintenance (%)
	Estimated cost percentage of the operations and maintenance function that is outsourced.
	Water monitoring services (%)
	Estimated cost percentage of the water monitoring services activities that are outsourced.

The level of management sophistication is also another type of explanatory factor that may be relevant to either the understanding of results or checking how much performance is achieved and "explained" by them. The characteristics of customer service (very important to interpret the complaints PI, for instance) and existing technological resources are two of the most significant examples, expressed respectively in Table 47 and Table 48.

Table 47. Other explanatory factors: customer service.

	Existence of system to record all customer complaints
	Existence of registers that record total number of verbal and written customer complaints, enabling nature of complaints to be determined by scrutinising individual entries.
	Existence of formalised system to record all customer complaints for service quality monitoring and asset management purposes
	Existence of customer complaints recording and data processing system that is used for resolving customer complaints, monitoring of service quality and performance and asset management planning.
	Existence of a guaranteed standards scheme
	Existence of a guaranteed standards scheme (GSS) that establishes the rights of the customers, including at least: minimum service pressure at the *delivery point*; maximum time to get a new connection and to *repair* an existing one; maximum time of written responses; appointment times to attend customers' premises. *The set times existing in the GSS should be specified.*
	Target time for a new connection
	Maximum time targeted for establishing a new connection, from the customer request until the availability of the service (in the cases when the service connection exists).
D39 –	Residential customer meter reading frequency (No./meter/year)
	Frequency of residential customer meter readings, pre-defined by the *water utility* for the assessment period.
D40 –	Industrial customer meter reading frequency (No./meter/year)
	Frequency of industrial customer readings meter, pre-defined by the *water utility* for the assessment period.
D41 –	Bulk customer meter reading frequency (No./meter/year)
	Frequency of bulk customer readings meter, pre-defined by the *water utility* for the assessment period.
	Age of residential customer meters (years)
	Average age of residential customer meters.

Table 47a. Other explanatory factors: customer service (contd.).

D34	Average service pressure (kPa)
	Average service pressure at the network *delivery points.* *Service pressure depends on time and location/elevation, and is a variable that many utilities do not know accurately. However, it is an import piece of information.*

Table 48. Other explanatory factors: technological resources.

Computerised Information Systems (IT)	
	Planning and decision
	Routine use of IT to support planning and decision. (Values: yes/no)
	Billing or accounting
	Routine use of IT to support billing or accounting. (Values: yes/no)
	Maintenance
	Routine use of IT to support maintenance activities. (Values: yes/no)
	Customer complaints
	Routine use of IT to support customer service quality and network performance appraisal. (Values: yes/no)
	Other
	Routine use of IT to support other relevant management activities. (Values: yes/no)
Automation and control	
	Pumping (%)
	Number of pumping stations with automatic control units/total number of pumping *stations* × 100.
	Treatment
	Routine use of automatic procedures to control treatment units. (Values: yes/no)

Table 48a. Other explanatory factors: technological resources (contd.).

	Central telemetry
	Availability and routine use of a central telemetry facility. (Values: yes/no)
	Central monitoring and control
	Availability and routine use of a central monitoring and control facility (SCADA). (Values: yes/no)
	Integrated control (%)
	Number of *integrated stations*/total number of *stations* × 100. (Values: yes/no)
Mapping	
	Updated mapping (%)
	Length of mapped mains with a capture tolerance compatible to a scale 400 p.p.m./total mains length × 100. *(400 p.p.m. = 1:2500)*
	Digital mapping (%)
	Length of digitised mains with a capture tolerance compatible to a scale 400 p.p.m./total mains length × 100. *(400 p.p.m. = 1:2500)*

6. DATA QUALITY

6.1 Introduction

If a concept included in the IWA performance indicators framework has failed to be widely embraced and applied is the actual integration of data quality in the performance assessment systems. Even today, very few systems in the world take data quality into consideration or manage it according to its importance. And yet, understanding, collecting and taking into consideration during the assessment period of the quality of the collected variables is critical to make coherent and well informed decisions.

The water industry is remarkably prone to deal with information of poor quality. With a few remarkable exceptions, the buried nature of most of the assets and the long period of time since the water services first initiated their activities often lead to unknowns and uncertainties. Information systems are often not completely up-to-date or not adequately connected to each other. Furthermore, although universal metering is a reality or a goal for most utilities in the world, metered volumes are often inaccurate due to old or non-calibrated meters. This reality often becomes evident when a utility chooses to participate in a performance assessment project to realize that data is often not readily available or of enough quality (and this usually comes as a surprise).

Performance indicators are most useful when compared (to either targets, previous performance or peer results); however, making decisions based on such comparisons when data quality is poor is a risky exercise that should be carried out with extreme care.

However, most uses of the IWA performance indicators found in practice present the numerical value of the indicator with no information on how good the quality of the data used to achieve such value is. This can create situations in which the indicators misinform instead of guiding to good decisions. This is particularly true in comparative performance assessment, where the values of indicators for several utilities are compared (for instance as used in yardstick competition by some regulators).

A good illustration of this problem has been often given by some of the authors in some public presentations for some time. Imagine two hypothetical utilities comparing their real losses performance. Table 49 summarises the information leading to the values of the real loss indicator for each utility. As shown in the table, the difference in the quality of the data leading to the value of the two indicators is great.

Table 49. Example of data quality differences in two hypothetical utilities.

Variable/Indicator	Utility A	Utility B
Input volume	Flow meters are present and maintained at the entry of the system	No flow meters. Input values are estimated from pumping times and estimated pump efficiency
Authorized consumption	All uses are metered, including residential users. A meter maintenance programme is present	No meters used for billing. All volumes are estimated
Number of connections	Each customer with a contract has a single connection. Certainty is high.	Number of connections is unknown. Organic growth has lead to cases of several connections per customer, and total value is unknown.
Real losses indicator	260 l/connection/day	215 l/connection/day

However, in most projects observed to date the results would be presented as:

- Utility A: 260 l/connection/day
- Utility B: 215 l/connection/day

Most water professionals when confronted with this example immediately understand the need for a systematic approach to data quality in performance indicators system. An approach that has always been a part of the IWA framework for performance assessment, but unfortunately still remains overlooked.

The authors have often wondered why the previous two versions of this manual failed to complement the wide acceptance of the proposed indicators with a similar enthusiasm in embracing data quality management. The answer has always pointed to complexity in some way or the other.

While the fundamental concepts introduced in the very first edition of this manual still remain valid and completely applicable, it is true that the proposed system has proven to be too complex to be applied under many circumstances.

6.2 Complete data quality management: Accuracy and reliability

The original framework to manage data quality published with the first IWA manual on performance indicators was based on the system developed by OFWAT for its regulatory scheme.

This system remains completely valid today and it provides detailed information on how good the data used to calculate the performance indicators is. However, as mentioned above, the users of the system have often found this option too complex.

To perform a full data quality assessment of the input data, it should be assessed in terms of the reliability of the source and of the accuracy of data.

The reliability of the source accounts for uncertainties in how reliable the source of the data may be. For instance, the extent to which the data source yields consistent, stable, and uniform results over repeated observations or measurements under the same conditions each time. Additionally, data in the water sector is often obsolete and based on incomplete records, making advisable to record how reliable the source of information is.

The accuracy accounts for measurement errors in the acquisition of input data; the closeness of observations, computations or estimates to the true value as accepted as being true. Accuracy relates to the exactness of the result, and should be distinguished from precision, which relates to the exactness of the operation by which the result was obtained.

Practice shows that, in general, data providers do not have detailed information on reliability and accuracy, but are able to provide informed guesses, if broad bands are adopted. The field test of the IWA PI system showed that the use of four categories of data accuracy and three for reliability is feasible and a good balance between "an ideal world" and not using this type of information.

Recommended accuracy bands are:

Table 50. Recommended accuracy bands.

Accuracy Band	Associated Uncertainty
0–5%	Better than or equal to +/–5%
5–20%	Worse than ±5%, but better than or equal to +/–20%
20–50%	Worse than ±20%, but better than or equal to +/–50%
>50	Worse than ±50%

Recommended bands for the reliability of the source are:

Table 51. Recommended data source reliability bands.

Reliability Band	Definition
★★★	Highly reliable data source: data based on sound records, procedures, investigations or analyses that are properly documented and recognised as the best available assessment methods.
★★	Fairly reliable data source: worse than ★★★, but better than ★.
★	Unreliable data source: data based on extrapolation from limited reliable samples or on informed guesses.

It is recommended that specific implementations define criteria to assess the reliability bands for every input variable (e.g., *** – variable assessed based on metering that contributes to at least 95% of the total value).

Data source reliability and data accuracy should be assessed for every input data element. Accuracy of the calculations should be assessed as well, preferably in a quantitative way, according to the ISO Guide to the expression of uncertainty in measurement[6]. See II-5 for details on how accuracy of values resulting from algebraic operations can be assessed.

6.3 A new focus on data quality

The complete approach to data quality management as described above, has proven to be fundamental to decision making when adequately applied, but also too complex for many projects. Especially those where data quality is quite poor.

As a consequence, some of the applications of the IWA system that have appeared in the latter years have implemented simplified data quality recording systems. Most of these systems share two basic characteristics:

[1] A single scale for data quality. Reliability and accuracy are replaced by more generic terms and data is qualified by terms like "good", "average" or "poor".

[2] Each data quality level is defined for all variables. Since data is qualified in more generic terms, it is necessary to define for each variable what is understood as "good" or "average" quality data.

This new approach requires greater effort when defining the performance assessment system, as a definition is required for each data quality. These definitions should be tailored to the specific project needs and the reality of the participants in the project. However, once the levels have been defined, utilities may find it much easier to provide an assessment of the quality of data.

[6] GUM (1995) – Guide to the expression of uncertainty in measurement, ISO, ISBN 92-67-10188-9.

7. IMPLEMENTATION OF A PI SYSTEM

7.1 Introduction

Performance indicators are a powerful management tool as has been demonstrated through their systematic use in many industries for decades. They can provide measures of how many resources are being used in relation to available ones, they can be used to assess the extent that management targets are met and even to evaluate the general impact of management strategies.

However, performance indicators cannot become a goal in themselves. The implementation of a performance indicators system must be the result of a wider approach in management within a *water utility*. Otherwise the results obtained through performance indicators may not be adequate or even useful at all. A typical consequence of performance indicators being used as the end instead of the means is described by Neely (2002) in the so called "*Measurement crisis*":

The measurement crisis

- The measures bore no relationship to the strategy
- The wrong things are measured
- Everything that is easy to measure is measured
- The only relevant measures are the financial ones
- Sub-optimization is encouraged
- Things are measured but not analyzed and no actions are taken as a result
- The operational level is overlooked
- The measurement systems are so complex that the priorities are not clear
- There is a good synthesis of past performance but poor future forecasting
- Everything that walks and moves is measured, but nothing that matters

Some of the problems described above are quite common in many water *utilities*, including some that have implemented a performance indicators system. Even the First Edition of this manual might have misled some readers in the use of performance indicators by suggesting that all indicators in the manual were to be used.

However, the intention of the authors of this manual is not to provide a single unique set of performance indicators, but rather to provide a large comprehensive set that would cover many of the needs of a typical *utility* in most parts of the world. A detailed study of the performance indicators contained in the IWA proposal clearly shows that some of them have been created to be used only in certain parts of the planet and under specific conditions. Consequently, an appropriate use of the system requires a selection of a subset of indicators and correct implementation of the system. This chapter is devoted to providing some basic guidelines for the correct use of a performance indicators system based on the IWA proposal.

7.2 Using IWA indicators in a performance measurement system

A performance measurement system is a necessary tool in any company and consequently in any *water utility*. A good system will provide information on the current performance of the company and may even anticipate future events. The ultimate aim of any performance assessment system is to provide information as opposed to data (we can define *information* as any data which is useful in a decision making process). Consequently the data collected and processed within a performance measurement system must be related to the goals and interests of the company and its managers.

A proper performance indicators system must be understood as part of a wider entity which links objectives, strategies, critical success factors within the company and the indicators themselves. This integration of performance assessment within all management activities is one of the foundations of systems such as the *Balance Scorecard* (Kaplan & Norton, 1996) which aim to show that the main perspectives (financial, customer, processes and learning and growth) within a company are interconnected and consequently must be considered as a whole.

The IWA system of performance indicators provides a very valuable tool when adopting performance measurement systems such as the Balance Scorecard, and both tools may be fully compatible. A successful integration of such a balanced system with IWA indicators mainly depends on the correct philosophy and approach and a careful and methodical implementation.

The implementation process described in this chapter is generic and does not correspond to a specific performance measurement system. However, a proper knowledge of any other system and the IWA performance indicators proposal should allow for a relatively easy integration of this proposal into any system.

7.3 Definition of the objectives

The implementation of any PI system has to be objective-oriented. Performance indicators are the last step of a larger management strategy that should link the *utility*'s objectives to strategies, define critical success factors and only then bring performance indicators both as means to evaluate the success of these strategies and as a control mechanism to detect problems in advance.

Objectives need to be precise and clear. They need to be both demanding and realistic and most importantly reflect the mission and vision of a company. Short term objectives should be part of a long term planning strategy, which is especially important in a heavily asset based business such as water supply. Objectives for a *water utility* should take into account not only the purely managerial aspects of a company, but also the needs and goals of the different *stakeholders*.

The definition of objectives should always be the first step of a well-defined performance measurement system. A clear definition of the objectives to be achieved will be useful in gaining a perspective of the project in order to answer some difficult questions: How much should be invested in the performance measurement system? Who and which departments are to be included in the project team? What will be the benefits of such a project?

Failure in properly defining the objectives will only lead to uncertainty in the usefulness of the whole system and lack of implication at all levels. The definition of these objectives can span from high priority objectives at company level to periodic objectives at departmental level. Depending on the importance of the objectives, the managerial staff ultimately responsible for those objectives should promote, encourage and clearly back up the whole project. In *utilities* with total quality management strategies already developed, the performance measurement system should be integrated and adopted as an additional tool.

Water supply is a public service and presents some differences when compared to other industries. The presence of external *stakeholders* (in addition to internal ones) may affect the definition of the objectives and ultimately the choice of performance indicators. In any case, the methodology to apply would remain the same and the only changes would be the nature of the objectives and the actors playing a role in the process.

A list of motives for using a PI system (and as a consequence of possible objectives) is presented in I-1.3, and illustrated in Figure 11.

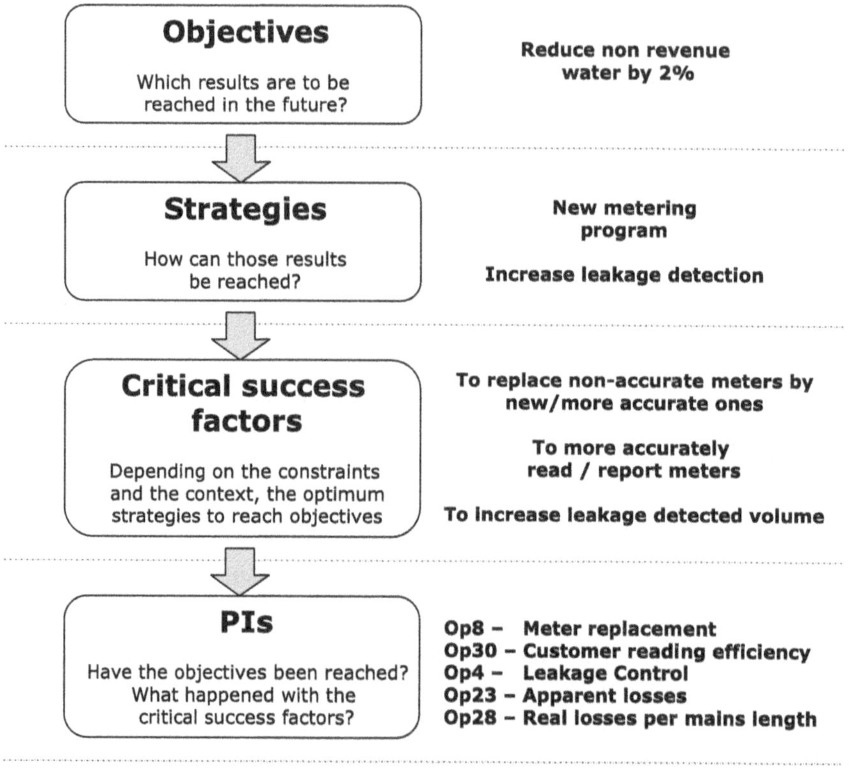

Figure 11. PI as a part of a performance measurement system.

7.4 Definition of strategies

The strategies that need to be applied in order to reach the objectives in a *utility* are a key management decision. This decision depends on many factors, especially in the water supply sector where variables not directly related to business criteria come into play (usually corresponding to the need for fulfilling the objectives of the external *stakeholders*).

This manual does not intend to provide potential strategies for managing water utilities. However, and regardless of the strategies chosen, they should be linked directly to one of the objectives established in the previous step. All the objectives defined should be addressed by at least a specific strategy and all the strategies developed should be the result of trying to achieve at least one of the objectives.

7.5 Establishment of the critical success factors

It is not always possible to reach objectives using all strategies. Quite often managers are faced with restrictions which may be as simple as economic ones, or may even present a political nature. Additionally, the context may deem a successful strategy inappropriate for a particular case. As a result, the selection of the optimum strategy to reach the utility's objectives will depend on the context and the existing constraints in each situation.

Critical success factors are as a consequence dynamic entities. While the previous phase relies on identifying all possible strategies that may lead to reaching the objectives, the selection of the actual critical success factor will depend on the time, the place and the situation. If the constraints change, or the context is different, this may lead to a different critical success factor. It is consequently of vital importance to track the evolution of all constraints present (something quite obvious for any utility) and to be aware of the context where the decision is made (and thus a strategy chosen for a certain utility may not be the best for another one with a different context).

7.6 Establishment of a performance indicators system

Performance indicators are the main aim in this manual. However they should always be taken into account as part of a larger picture. Performance indicators play the role of controlling, assessing performance, identifying success in reaching targets and providing valuable information for forecasting future trends and planning needs.

There are many definitions of sets of performance indicators in the water industry and outside the water industry. All of them can be a useful source and starting point for an adequate selection of PI in a system. However not all of these sets have been defined with the same level of detail or the same objectives in mind.

The IWA system of performance indicators for water supply was developed taking into account three key factors:

- It was developed for the industry with internal management objectives in mind;
- It was developed trying to take into account the different realities of water supply around the world, and aiming to provide valid indicators for all those realities;
- It was designed to allow for truer comparisons whenever possible, in order to be useful for comparative performance assessment.

As a consequence, the IWA WS PI system is a large set of indicators that needs to be understood in its context. It presents a large number of advantages over other existing systems, and a few disadvantages. Users should bear these strengths and weaknesses in mind when using the system.

Advantages:

- Solid structure – As defined in chapter I-3, the IWA PI system has been systematically built around a coherent, clear and solid structure that allows for modifications, adaptations and enlargements.
- Thoroughly defined system – Strong definitions of the different elements (indicators, variables, context information) are a most valuable starting point for the user. The system has been tested in hundreds of situations and in every part of the world and the result is a set of definitions that take into account most of the issues that may arise in the use of the system.
    - Well balanced set – The IWA system was developed taking the *utility* as a whole. As a consequence the number of indicators devoted to each part of the service has been designed to be kept in balance with the rest. The use of the IWA system as a whole should provide an equal amount of information on every aspect of the *utility*.
- Standard definitions – The IWA system has already been considered a standard and has been taken into consideration worldwide when new systems have been defined. By using the indicators defined in this manual, the chances are that someone, somewhere will be using the same definitions.

Disadvantages:

- Too shallow – The set of IWA indicators was designed without a specific need in mind, but rather for looking at the service as a whole. This must result in a lack of some necessary performance indicators for more specific applications.
- Too broad a scope – IWA indicators were selected in such a way that at least some of them could be used to assess water supply services performance anywhere in the world. Since the reality of water services is so different around the planet, parts of the system may appear useless when considered for some *utilities*.
- Too many indicators – As a direct consequence of the two previous points, the number of performance indicators presented in the IWA proposal seems too large, and it certainly is for the average application and *utility*. However, the large number of indicators also offers more possibilities for choosing the most appropriate ones for every case and circumstance.

As shown above, the IWA system is not all good news to start with. However, all the problems mentioned before can be easily overcome through a proper establishment of the system. Specific needs may be addressed by adding or modifying indicators and a proper selection process will discard useless or meaningless PI for the actual needs of the user. Regardless of the changes made to the system, the process will be much easier and will result in a more solid set than the one that could be obtained if built from scratch.

It is consequently of great importance to follow a methodical implementation of the Performance Indicators System. Such steps are described at this point.

Definition of the strategic performance assessment policy
The definition of a performance assessment policy as such is the result of applying the concepts defined earlier for any performance measurement system. In any case, it should be considered at any level, the first step for any performance indicators system. A clear policy with defined objectives and strategies, and with specific critical success factors will be easier to address than broader topics or notions. These decisions are the responsibility of top management, and a **strategic team** should be in charge of working out the strategic direction that the utility should take in the future.

In this phase, the utility should ensure that the objectives include those of all internal and external stakeholders. In this way, those commissioning and driving the process, those carrying it out, those providing the key data and those interested in the outcomes are effectively engaged.

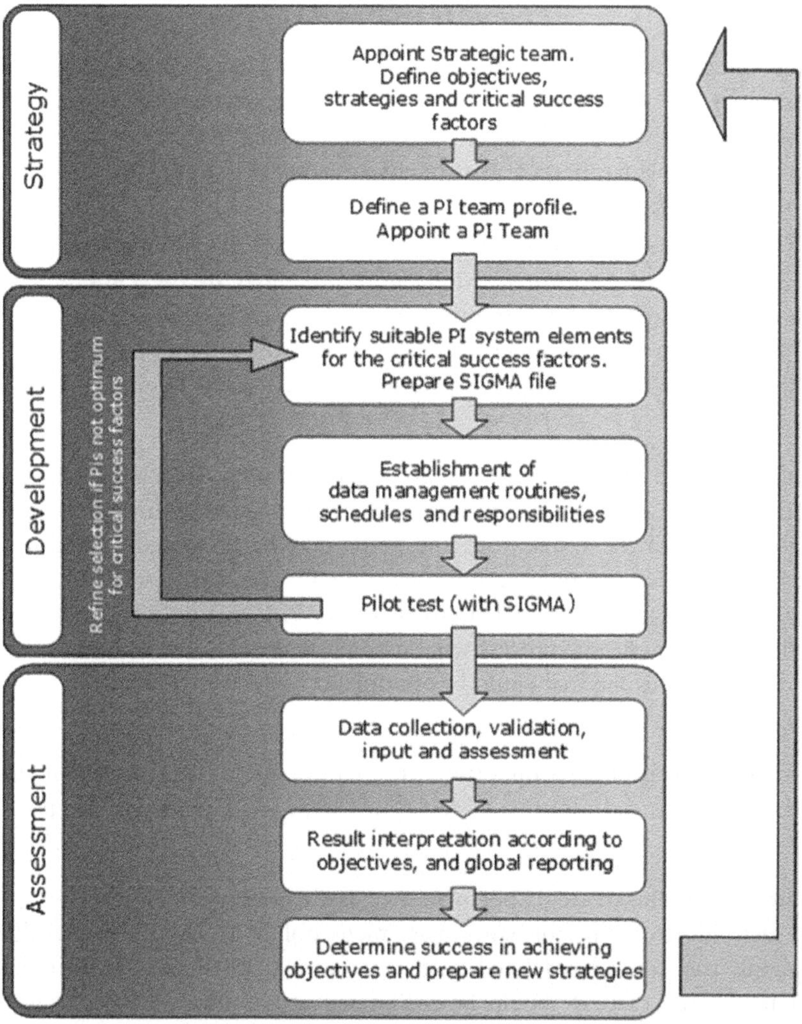

Figure 12. Phases of the PI system implementation process.

This has proven to be a good starting point to determine the characteristics and constituents of the **PI team**. This team should be in charge of the actual PI project and its consistency within the strategic decisions previously made. As a result, the successful implementation of a PI system within an organisation normally requires that the PI team leader is a senior staff member, if necessary assisted by less experienced colleagues. The involvement and active participation of the senior staff, usually the decision-makers who are going to use the PI information, is essential. It is also strongly recommended that representatives of the departments/sections who need to provide the PI data are core members of the PI team, in order to ensure their commitment to all the stages of the process.

Experience from pilot testing the IWA PI set has demonstrated that an important benefit of the implementation of a PI system may be derived from staff participation. Participants experience a wider range of alternative views and perspectives from other staff members with different responsibilities and backgrounds. It is essential at this stage to properly brief the PI team about the purpose, objectives and expected outcomes from the establishment of the performance assessment system within the *utility*.

The next stage in the process is to review available PI systems and select the most appropriate to the specified critical success factors. Although this manual may be used to generally assist users in the application of a wide range of PI systems, it is assumed that users will ordinarily adopt the IWA Manual of *Performance Indicators for Wastewater Services*. Subsequent guidance in this manual may not be applicable to other PI systems.

Selection of the PI to be assessed

The next step in the process of establishing a proper PI system consists of adequately selecting and defining the Performance Indicators that will be used and the necessary Context Information (CI) in case it is needed. The selection stage becomes relevant for several reasons:

- The total number of performance indicators may be a key factor in the success of the project. Too many indicators and the costs and difficulties of implementing the system will increase significantly. Too few and the system will not be able to provide a proper assessment of the performance of the *utility* in the terms defined by the objectives and the selected strategies.
- The indicators selected must provide a balance in studying the different areas of the business. Selecting a small number of indicators is not of much

use if all of them refer to the same area of the business (and the objectives and strategies refer to a wider scope).

- The nature of the indicators selected may influence the interpretation of the results. Although a properly defined indicator will provide little or no bias to the interpretation of its results, there are many real life examples which show us that this is a real danger. For instance, it is a well-known fact that the volume of water losses expressed as a percentage may be a misleading indicator in certain circumstances.
- Indicators need to be consistent in their construction. A well-defined PI system will consist of performance indicators that share the same structure and level of detail in their definitions, assessment period and units. Additionally the system should be built in such a way that the common variables are shared by indicators and have unique definitions, assessment periods and units.

Taking all these factors into account it is not difficult to see why the IWA proposal provides a powerful tool for building a PI system. When using the IWA proposal the only difficulty that needs to be overcome is to select a reduced number of PI depending on the needs of the project. All the other conditions, which are time consuming and require experience in performance measurement, are already fulfilled.

However, this does not mean that additional indicators cannot be used. As a matter of fact, in a good number of cases it will be necessary to create and add more indicators to the selected subset of IWA indicators. These new indicators will take into account the peculiarities of the region, the business, the stakeholders or the utility itself. However, all new indicators should be built in such a way that makes them fully consistent and compatible with the IWA indicators.

In order to select the appropriate indicators and context information from the IWA set, the following five steps are presented:

[1] Pre-selection of the PI considered to be of importance for the assessment (and relevant to the selected objectives, strategies and critical success factors).

[2] Selection of the actual PI and CI to be used in the initial assessment.

[3] Development of additional PI and CI if necessary. Consistency check.

[4] Pilot testing of the initially selected PI.

[5] Final selection of PI deemed to be important and to be used in the assessment.

These processes are illustrated in Figure 13.

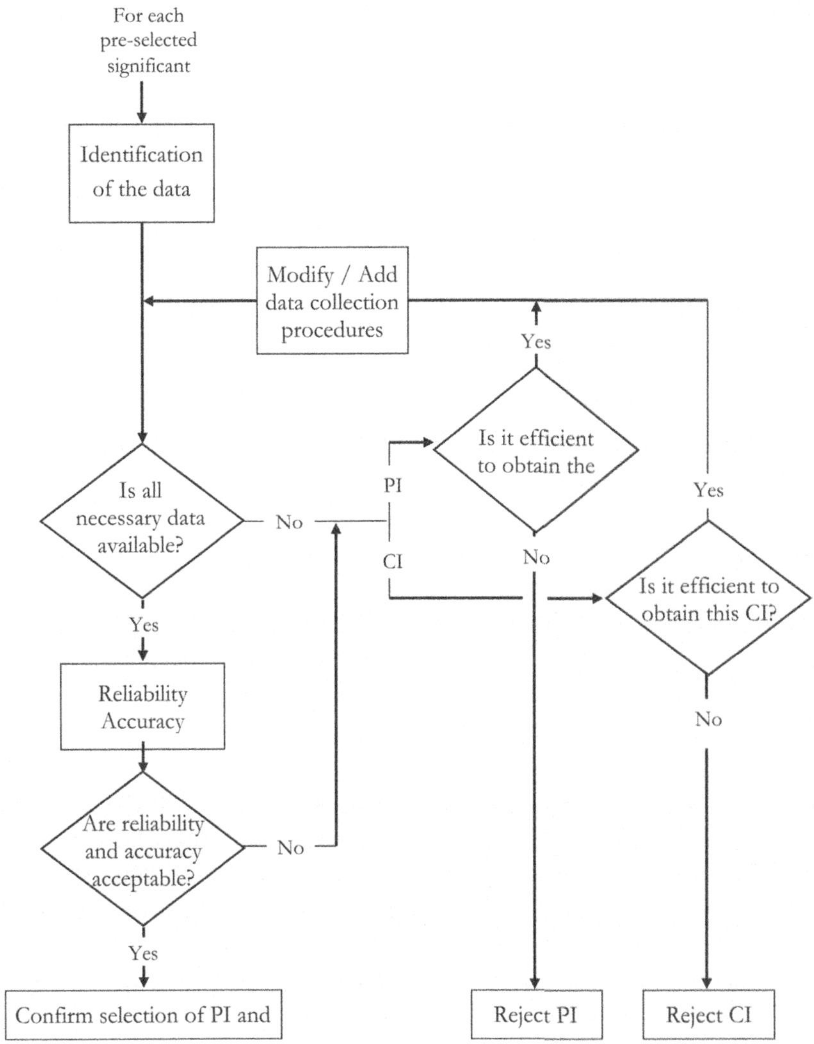

Figure 13. PI and CI selection procedure.

The first step is to classify the indicators depending on their importance for the assessment. This classification should take into account the views and needs of the several stakeholders involved. A three level classification (*high, medium* and *low*) may be appropriate for this stage. This ranking should be carried out regardless of the availability or quality of the data needed for the PI. Once the important PI have been identified it may be necessary to modify data collection and handling processes, and these improvements should be done taking into account the importance of the indicators that are to be calculated from these data.

The initial selection of PI should be made at senior level (Strategic team) within the utility in consultation with the PI team, who would interact with the suppliers of the data (at the department level). Additional consultation outside the utility may be required where, for example, the performance assessment is being conducted for regulatory or external purposes. An illustration of the participants that may be required in assigning levels of significance and selecting relevant PI is given in Figure 14. External stakeholders may not be consulted at all when the evaluation is for internal use.

Should there be a PI of high significance, but with unavailable or difficult to obtain variables, the PI should still be included in the selection and measures should be taken to obtain the data for future PI assessments.

For some assessment needs it may be necessary to customise the core PI. These may need to be supplemented with expanded or more detailed sub-indicators. The selection of these other indicators should follow the same procedure. Following PI pre-selection, it is recommended that an individual, a department or a sub-group is assigned to be responsible for the data provision, validation and confirmation for each PI.

Following assignment of significance levels to each of the PI, the next stage is to pre-select the PI to be used in a pilot assessment. The selection should include those rated as most important. The final selection of indicators to be used after full implementation will be the result of an iterative process. In any case, the selection should aim to include the indicators that are optimum to track the critical success factors. However, sometimes the required data is either not available or not good enough. In such cases, the question of whether it is efficient to collect such data should be asked before proceeding to collect it or selecting another indicator.

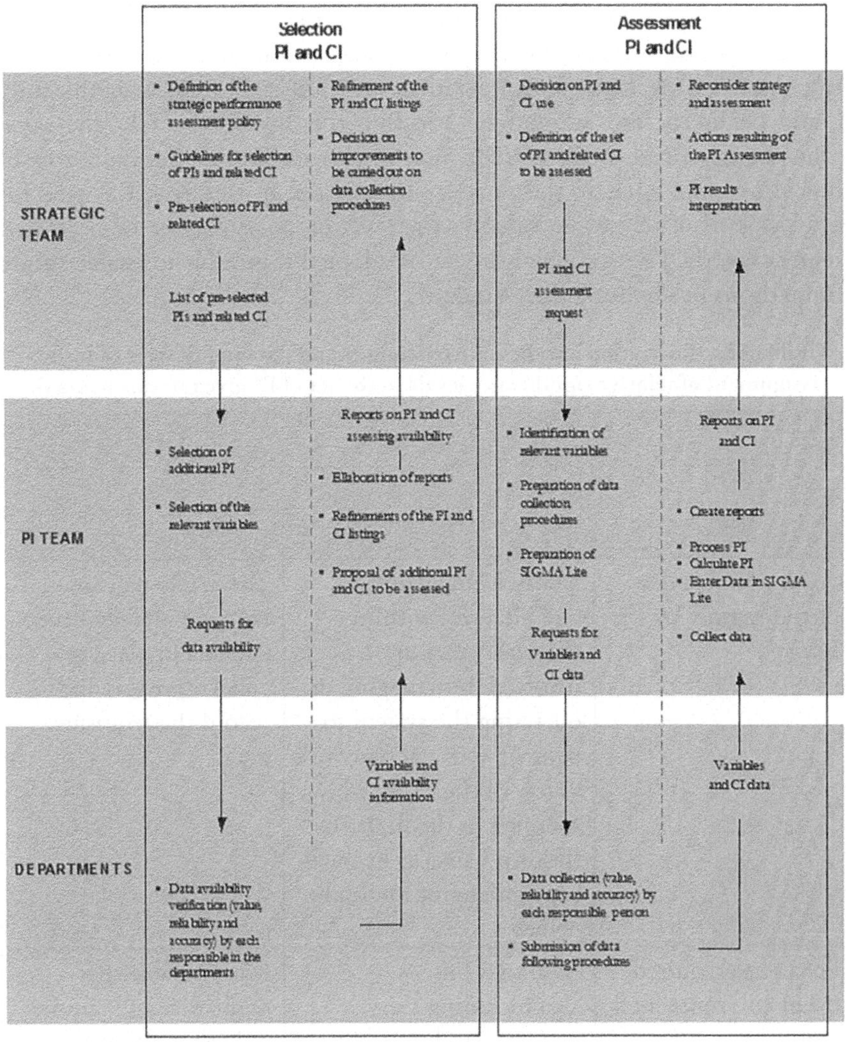

Figure 14. Example of data flows concerning PI and CI and team responsibilities.

The experience of the senior staff at the strategic team should provide the initial selection giving a preliminary listing based on the significance level. Then the definition of the data required and its confidence grade should be taken into account when evaluating the adequacy of each PI.

It is important to point out that in the case of preparing a performance assessment project, the Context Information should also be selected at this stage. The selection of CI depends not only on the local characteristics of all the participating *utilities*, but also on the actual differences between them. Any suspected element of context that may affect PI values significantly should be taken into account by selecting the appropriate CI. The same principles described in the introduction of this section for the indicators of the IWA system apply to the context information. In this case, the need for new elements of CI not included in the IWA proposal may be even greater. In any case, the IWA proposal should be adopted if possible to enable future comparisons by choosing the "standard".

Table 52. Illustration of types of participants and data implications for the assignment of relative significance levels to the set of PI given in this manual.

Example of the purpose (context) for using PI system	Stakeholders involved in assignment of relative significance to each PI	Data need implications
For internal review purposes only by the *utility*	Senior Management and all levels within the *utility* that are involved in managing or operating the system to be reviewed. As many or as few of the PI may be assigned to the highest (priority) level as deemed appropriate for internal review.	The *utility* may or may not already have systems in place to collect, process and record the requisite data.
For use as required by an environmental regulator	As above, but also including the regulator and possibly proactive users, such as environmental interest groups (NGOs). It is probable that there will be some negotiation as to the PI to be considered at the highest level of significance.	It is probable that regulators will require only those PI to be assessed that are relevant to their individual area of interest, e.g., economic or environmental. It is likely that the *utility* will already be recording the required data.

Table 52. Illustration of types of participants and data implications for the assignment of relative significance levels to the set of PI given in this manual (contd.).

Example of the purpose (context) for using PI system	Stakeholders involved in assignment of relative significance to each PI	Data need implications
For interaction with a wide range of customer groups	As wide a range of stakeholders as possible. The *utility* will have first to convince the other stakeholders of the need to prioritise the PI and report only on those most significant, but it may be necessary to eventually increase the number of PI reported.	It is unlikely that the *utility* will have all of the data needed for the entire PI perceived to be significant by diverse groups of stakeholders. A more limited set of indicators may already be in place as part of a customer 'council' or other form of representation.

By working successively through the pre-selected, each one can be considered for inclusion or rejection for the final list of the most important PI according to the criteria previously established. The pilot test may provide additional information on the adequacy of the indicators to track the critical success factors, and whether these indicators are efficient in doing such jobs. Consequently, it is possible that PI initially ranked as medium or low significance, may in fact be found to be of greater importance given the circumstances.

The PI implementation team has to identify all the data required and work with the data providers in order to assess data availability, reliability and accuracy. Whenever data for a PI is not available or is inaccurate, the strategic and PI teams and the data providers should decide whether or not to improve the data or to select another suitable indicator. An important element of the decision will be the possibility of linking current information systems with the performance assessment system. Optimally, the data in all information systems within the utility should be consistently defined, collected and managed in order to maximise the benefits derived from their management.

Figure 14 shows how the various participants in the PI implementation process interact, together with data flows.

7.7 Assessment of the performance indicators

Assessment of the values of performance indicators of the system should be done accordingly to the objectives, strategies and critical success factors established beforehand. A performance indicator cannot be assessed unless it is compared to a reference value. These reference values can be predefined objectives, past values of the same performance indicator or external values of the indicator (obtained from benchmarking projects or external references). The process of result interpretation usually requires comparison with a reference value or a reference threshold, taking into account the context information, and analysing a group of related PI rather than an individual indicator.

In order to maximise the benefits of the assessment, a systematised approach should be used. The assessment procedures should be defined by the Strategic and PI teams, and all participants in the process should receive some sort of feedback from the PI assessment in order to maintain the motivation in the project (from the data collectors to the top management).

The results from PI assessment can be produced using the SIGMA software that can be downloaded from http://www.sigmalite.com/iwamanual/, in combination with other spreadsheet software packages. Alternatively users may utilise individual databases. Final reporting should be customised to account for the intended use of the PI information. It is important that the relevant context information is included as well as the PI. Reporting and evaluation of the outcomes of the assessment are considered in Section I-8.

The immediate consequence of assessing the results provided by the PI should be an improved decision making process. In the first stage, performance indicators should provide information about the adequacy and success of the strategies defined as critical success factors in achieving the objectives. As a result, performance indicators results may be used to make immediate decisions or to define long term strategies.

Finally, it is recommended that upon completion of the PI assessment, the PI team and senior managers review the process, aiming to improve both the overall procedure and the detailed stages. In many cases an utility may wish to customise the process to correspond more closely with other auditing procedures. Where external comparisons are to be made for benchmarking purposes, this may also require some reconsideration of the PI included in the assessment.

Where PI are intended for use in comparative performance assessment, it may be necessary to compare PI with those from other parts of the *utility* performing similar functions, with other organisations, with pre-set targets for financial or environmental compliance, or other benchmark standards. In many cases the PI exercise will serve more than one of these purposes.

Where cross-organisational benchmarking is carried out, individual *utilities* may wish to safeguard the confidentiality of some of their PI. Under these circumstances comparisons of PI can be made anonymously within carefully selected sub-groups, or managed by an independent external entity that collects the PI data on behalf of the various *utilities*, and then processes and reports the information in a form that ensures that individual undertaker identification is not possible.

7.8 Continuous improvement

All the processes presented in this chapter are intended to be part of a continuous improvement process within a utility. It is quite obvious that defining objectives and strategies to reach them is a periodic task in the management of any organisation at any level.

Performance indicators are not a goal in themselves, although they can play a key role in successfully leading a utility to more efficient management and a systematic approach to face and achieve targets. However, the benefits of the implementation of a performance measurement system using indicators will increase with time as the system is refined and assimilated in the organization. Quite often utilities are put back by an initially unsuccessful application of performance indicators and hence giving up on the project by not achieving water service results.

For these reasons, the implementation of a PI system in a utility should be considered a mid-term project. If only limited resources can be allocated to the project, it could prove more interesting to reduce the size of the system and expanding the life of the project, rather than aiming for a full size PI system that will only be used once. Experience makes a difference, and performance indicators are a clear example.

8. EXAMPLES OF APPLICATION

8.1 Introduction

The objective of this chapter is to illustrate the application of the IWA PI system for water supply services in different instances, in order to demonstrate its flexibility and some of the ways it can be used. Each example aims to be a source of inspiration, and should not be taken as a guideline for similar cases, because assumptions and decisions are made that may not be appropriate if the circumstances are different. The responsibility of the texts are of their respective authors and do not aim to reflect any consensus among IWA members.

The first example reports the use of the IWA PI system within the scope of regulation. Multiple examples can be found in the world. The first one, and therefore with a longer record of application, is in Portugal. An objective-oriented PI system was implemented in 2004. Almost all of the PIs of this regulatory assessment system are IWA PIs. The example from China aims at illustrating the possibility of a large-scale implementation of a PI system, aiming at supporting several levels of administration, as well as the individual utilities. The case of Aquarating demonstrates that PI systems may be part of broader management assessment systems, where the understanding of the management processes can – and should – be associated with their efficiency and effectiveness assessment. The case from Germany shows how a nationally harmonised PI set was tailored, based on selection and modification of the existing IWA PIs. Last but not least, the case of the application of PIs to infrastructure asset management aims at demonstrating and illustrating how a performance assessment system is instrumental to any core organizational process. PI systems, in some cases evolving to performance assessment systems that include not only PIs but also other types of performance, risk and cost metrics, allow to formulate clear objectives and targets, support diagnosis, compare alternative paths, monitor results and identify deviations.

All the cases presented use the IWA PI systems as a basis, demonstrating their broad applicability and level of implementation in the world. In most cases, some tailor-made PIs were used as well to respond to specific needs, context and assessment criteria. In all the cases the new PIs were designed following the IWA specifications presented in 2.2.

8.2 Case 1: Leverage effect of the regulatory quality of service assessment system in Portugal

Authors: Paula Freixial, Ana Rita Ramos, Francisco Mira, Maria João Moinante, Maria José Franco, Margarida Monte, Patrícia Silvério and Rute Rodrigues

Country/region	Portugal
Scope	The water, wastewater and waste management services
Objective	Regulation
Leading partner(s)	ERSAR – The Water and Waste Services Regulation Authority
Participating utilities	The total number of utilities under evaluation is 295, with 274 utilities providing drinking water supply services and 283 utilities providing wastewater management services.
Character of the participation	Required by law
Public availability of results	Yes
Data audits	External audits promoted by ERSAR

Context and background

ERSAR is the Portuguese regulation authority for drinking water supply, wastewater management and municipal waste management services and the national authority for drinking water quality. It aims to ensure adequate protection for consumers and users of water supply, wastewater and waste services by promoting the quality of service rendered by the utilities and guaranteeing socially acceptable pricing, materialized in the following principles: essentiality, indispensability, universality, equity, reliability and cost efficiency associated with the quality of service. ERSAR is committed to ensure that there is equity and transparency in access to the water and waste services and their operation and respective contractual relationships, as well as consolidating an effective public right to general information regarding the sector and each operators. It also contributes to the implementation of government policies.

According to its organic law, ERSAR's regulation model is composed of three action plans: on a first level, the structural regulation of the sector, which consists in contributing to a better organization and to the clarification of the regulatory rules; on a second level, the regulation of the utilities' behaviour, monitoring throughout the life cycle the utility compliance with legal and contractual requirements, economic issues, the quality of service, the quality of drinking water and the interface with consumers; on a third level, auxiliary regulatory activities, including the preparation and the regular dissemination of information and technical support to the utilities.

The regulation of the quality of service aims to improve the effectiveness and efficiency of the water and waste services. Hence, it helps to ensure compliance with public service obligations, in terms of universal access to services, adequacy of the quantity, quality and continuity of services and efficiency of the utilities. This regulation is done by assessing the performance of the various utilities following a set of indicators related to each service, allowing ERSAR to regulate by benchmarking. The benchmarking process enables the establishment of baselines and the definition of best practices, creating an artificial competitive environment within the sector.

In this sense, in partnership with the National Civil Engineering Laboratory (LNEC), ERSAR has developed one set of 20 performance indicators (PI) for each service wich was applied for seven years since 2004, only to the concessions subject to regulation – "1st Generation of the quality of service assessment system".

Since 2011 and also with the technical support of LNEC, ERSAR introduced the "2nd Generation of quality of service assessment system" to all water utilities in Portugal. This second generation, composed of 16 PI for each service, has evolved from the experience gathered during the eight years application of the first generation of PI. Both generations were inspired by the performance indicators published by IWA regardingthe water supply and wastewater management services.

Assessment process

The regulatory model applied includes the use of methods for evaluating the quality of service provided to users by the utilities and the comparison of the results with other similar utilities operating in different geographical areas). Both results are subject to public exposure, as it is considered that this practice encourages the operators to be more efficient and naturally achieve a better position.

Three essential areas are considered for an improved performance of water supply infrastructures:

- user interface;
- service management sustainability;
- environmental sustainability.

Currently, the PI used for drinking water supply services assessment are:

WATER INDICATORS		
	Protection of user interests	**Accessibility of service for users** AA01 - Physical accessibility of the service AA02 - Affordability of the service **Quality of service provided to users** AA03 - Service interruptions AA04 – Safe water AA05 - Reply to written suggestions and complaints
	Operator sustainability	**Economic sustainability** AA06 – Cost recovery ratio AA07 - Connection to the service AA08 - Non-revenue water **Infrastructural sustainability** AA09 - Adequacy of treatment capacity AA10 - Mains rehabilitation AA11 - Mains failures **Physical productivity of human resources** AA12 - Adequacy of human resources
	Environmental sustainability	**Efficient use of environmental resources** AA13 – Real water losses AA14 - Fulfilment of the water abstraction licensing AA15 – Standardised energy consumption **Efficiency in pollution prevention** AA16 - Sludge disposal

Figure 15. Indicators of the assessment system for the quality of the service.

The quality of service indicators are defined in the assessment guidelines document (Alegre *et al.* 2013) and are the core of this system, allowing to measure progress over time and to evaluate if the main objectives of the service are being achieved by the utilities.

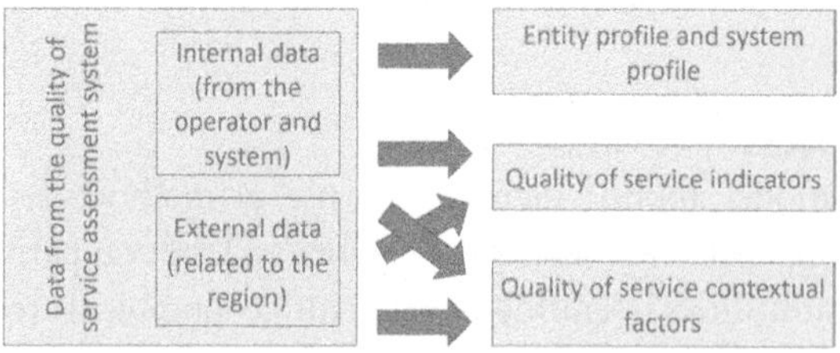

Figure 16. Components of the assessment system for the quality of the service.

In addition to the quality of service indicators, the system also includes information to support the interpretation of findings. Such information comprises the utility's profile, the system profile, other contextual factors not included in previous profiles and the source data that feeds this information. Figure 16 shows schematically the system's components and data flows (which totalized about 60 data for each of the services regulated).

The methodology applied by ERSAR is based on a sequence of steps in order to guarantee a clear, rational and transparent system. The annual procedure required to implement this methodology, from data collection by the operator, report to ERSAR by a digital platform ("Portal da ERSAR") to the disclosure of the final results by ERSAR, is the following:

OPERATOR TASKS	**Operator supplies data** ▪ Data collecting ▪ Self assessment of the quality of service ▪ Data input in quality of service module in ERSAR website ▪ Context factors selection (when applicable)
ERSAR TASKS	**ERSAR validates data for all operators** ▪ Compilation and cross validation of data ▪ Clarification of doubts ▪ Audits
	ERSAR processes data and interprets the results for each operator ▪ Analysis of indicators time evolution ▪ Interpretation of the indicators ▪ Contradictory procedure with each operator ▪ Indicators consolidation
	ERSAR processes data and interprets data for operators clusters ▪ Aggregation of the operators in clusters ▪ Summary of the results per indicator for each cluster ▪ Comparative analyses of the indicators per cluster of operators
	Publication and public disclosure of the quality of service annual report ▪ Preparation of the quality of service annual report ▪ Sending the information to each operator ▪ Public disclosure of the annual report

Figure 17. Annual assessment system of quality of the service.

ERSAR processes the data and proceeds to interpret it through the calculation of the PI. To each PI evaluation is given one of three colours (red, yellow or green), with resemblance to the traffic lights, according to the defined reference values and the existing contextual factors.

Summary of the performace assessment system

The performance assessment system implemented by ERSAR is summarised in Figure 18.

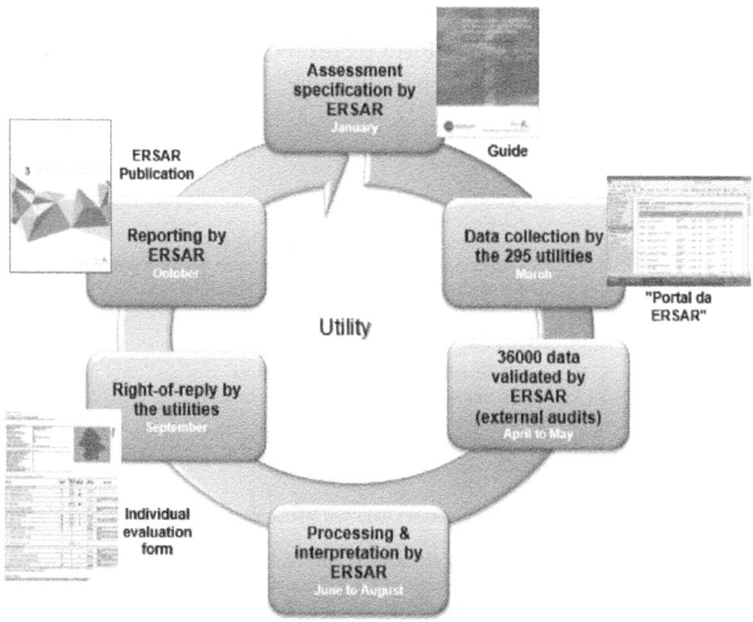

Figure 18. Annual assessment system of the quality of the service.

Results and learnings

Aware of the need for more reliable information about the sector, the implementation of the assessment system allowed ERSAR to start the publication of the Annual Report on Water and Waste Services in Portugal – RASARP in 2004, bridging this information gap.

The goal of the assessment system is to raise the utilities' performance to the highest possible levels, corresponding to a desirable "stage of excellence".

Over these 11 years, and despite the Portuguese assessment system is based on the IWA PI system, ERSAR has been making adjustments to the indicators, due to the extension of the scope of regulation (the number of utilities evaluated has raised from about 60 to about 300 utilities), the evolution of the water sector in Portugal and the diversity of management models, dimension and operation area of the utilities.

Finally, the need for upgrade also results from the fact that the IWA indicators are defined for the utilities management and the application of the indicator system by ERSAR aim to regulate the quality of service to improve the efficiency and effectiveness of the water and waste services.

The balance of the first four years (2011–2014) of universal implementation of the assessment system is very positive, highlighting the operators embracement to it. It is now accepted by the sector that the quality of service indicators, in addition to their regulatory objectives, are a particularly relevant tool to support systems management and to promote the ongoing improvement of the quality of service provided to users.

Regarding the results from the four first years of implementation of the 2nd generation, it was denoted that the utilities have an average response capacity of about 80–90% of the applicable indicators, although with large variation in answering capability and in the quality of data reported.

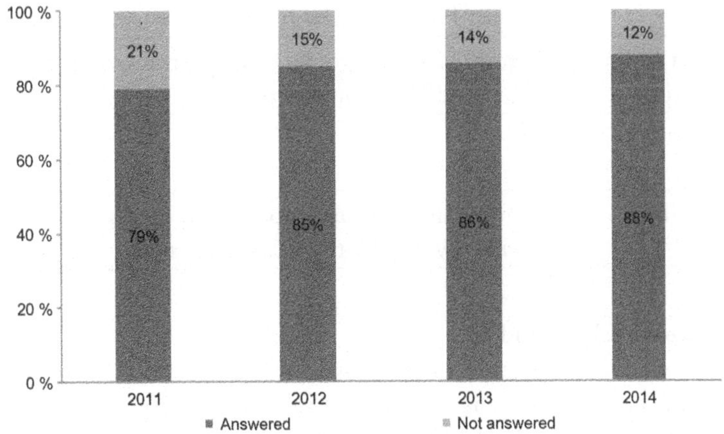

Figure 19. Percentage of responses in the first four years of application of the ERSAR assessment system – retail utilities.

8.3 Case 2: Performance assessment system of urban water supply in China

Authors: Han Wei, Li Shuang, Beijing Capital, Co., China

This case shows how a performance indicators system launched by the Ministry of Housing and Urban-Rural Development of the People's Republic of China for the urban water supply sevices was developed based on the IWA PI system.

Country/Region	China
Scope	Water supply services in urban-rural area.
Objective	To ultimate goal if to promote the sound development of urban water industry in China. The final purpose is to set up an urban water supply performance assessment management system which is appropriate for administrative departments of urban water supply of China' governments at various levels to monitor evaluate and check urban water supply enterprise performance.
Leading partner(s)	Beijing Capital Co., Ltd
Participating utilities	Development and testing stage: 6 water supply companies involved, which spreaded in Shanghai, Chengdu, Anqing, Maanshan Huainan, and Tongling cities. Implementation and extending study: in 7 cities of Jiangsu Province, which integrated urban-rural areas.
Character of the participation	Agreed with the provinces where the system has been tested and implemented so far
Public availability of results	Yes
Data audits	External and internal (up to now, carried out internatlly and by the BC project team)

Context and background

Research and Demonstration of Performance Assessment System of Urban Water Supply is a management subject (Subject No. 2009ZX07419-006), under the topic "Research and Demonstration of Drinking Water Safety and Security Technology" of the 11th Five-Year Plan of China National science and technology. It started in 2009 and was fully successful in the check and accept process in October of 2012. The implemtation started in 2013 in the Jiangsu Province. This subject was set up by Water Project Office of Ministry of Housing and Urban-Rural Development of the People's Republic of China (MOHURD), is coordinated by China Urban Water Association (CUWA) and is led by Beijing Capital Co., Ltd, which is a large state-owned enterprise in China's water industry. The participants include two universities, six water companies and a technology industrialization promotion agency.

The research team studied the most popular international performance assessment management system in detail and concluded that the IWA PI system was the most adequate to be used as the basis, in view of China's national conditions, and with the adequate adaptations.

IWA PIs started by being classified, evaluated and selected and adjusted or complemented when necessary, in order to respond to the requirements of various levels of government, industries and water utilities.

A pilot test and implementation phase followed in six utilities of approximately the same size but of different types and characteristics. Some are part of the Beijing Capital Group, but some are not. The implementation challenges are easier to overcome in the BC companies, because they share governance models, and have more processes and information systems in common. The test outside this universe was therefore very important.

Taking benefit from Cloud technology, a web-based platform was developed for data collection, preliminary validation and centralised analysis.

In the testing and validation period until 2012, this platform and of a manual collection system were used to collect the data from the utilities and results compared. In both cases the platform was used to archive all the data and allow for data analysis, followed field investigation and research to sort out the problems encountered.

Performance assessment process

One of the results of the first stage of the project (2009–2012) was the production of an implementation manual, where the PI system is specified and the guidelides for whole process. The PI system is composed of 49 variables and 24PIs.

The performance assessment is to be carried out annually. This manual defines the tasks over this annual cycle, and establishes roles and responsibilities for every task.

Special care is put on data quality control. An auditing process is defined. The data quality is specified based on the IWA accuracy and reliability bands.

The key stages of the annual cycle are:

[1] Data collection and upload by the utilities (a preliminary automatic validation is carried out when uploading the data).

[2] Data validation is a process to check data resource and identify data reliability and accuracy, which is important for data quality control.

[3] Performance assessment score including quantitative assessment based on PIs and expert assessment based on site investigation, each with its corresponding weight.

Final reporting publication and dissemination of the results. These reports are planned to be publicly available after feedback from utilities has been received. At study stage, the report for counterpart exchanges

8.4 Case 3: AQUARATING: An international standard for assessing water and wastewater services

(Authors: Enrique Cabrera, Francisco Cubillo)

Country/Region	Worldwide
Scope	Urban water and wastewater services
Objective	Standardised assessment of urban water and wastewater services
Leading partner(s)	International Water Association/Inter-American Development Bank
Participating utilities	Open to any utility
Character of the participation	Volunteer
Public availability of results	No
Data audits	External

In 2008, the Inter-American Development Bank (IDB) initiated contacts with the International Water Association (IWA) for the development of a system to increase efficiency and transparency for water utilities. After an extremely thorough development process, the result was AquaRating.

The system is managed through the AquaRating entity, a subsidiary of IWA, who received a license from the IDB to operate the system. This constitutes the first time that IWA itself operates a performance assessment system since the work on performance indicators was developed within the association.

The system relies on a series of performance indicators developed according to IWA guidelines (and some of them coincident with those found in this manual). Additionally, the system assesses the service through a series of yes/no questions regarding good management practices. Finally, AquaRating is specifically focused on reliability of information, data quality and its management. The system requires an external audit of data, and data quality itself has a significant impact on the resulting rating.

AquaRating was developed as a universal system to assess urban water and wastewater services. As a result, the scope of the system is retail water and wastewater services of a certain size. Depending on the characteristics of the assessed service, AquaRating is configured in one of the 16 possible configurations, excluding assessment elements that would not apply to the service (e.g., all wastewater indicators would not apply to a water-supply-only utility).

The system is not designed to compare values between different services, although statistical values from the sector might be obtained from it. It is therefore an individual effort, in which a service is assessed in a 6-year cycle. After a first assessment, with its corresponding audit, an AquaRating certificate is issued for the service. This AquaRating needs to be renewed every 2 years, although the second and third ratings do not require a full audit.

Performance assessment process

The performance assessment principles in AquaRating are based on those developed by IWA in its performance indicators manuals of best practices. However, the objectives laid out at the design stage of the project made it quite clear that it would be difficult to achieve the goals with the sole use of PIs. Specifically, AquaRating aimed to be a system that would:

- Allow a universal assessment of urban water and wastewater services around the world
- Have its data quality externally checked by a third party
- Provide a unique scale to measure all services around the world
- Provide comprehensive assessment across the services (including operational and financial matters, but also on concepts such as governance and access to service).

As a result, the team in charge of developing the system decided to base the assessment on three different aspects:

[1] If a criterion could be expected to be consistently measured around the world and the influence of context on the metric manageable (not having a very strong influence) then a performance indicator was assigned as an assessment element for such criterion. Some exceptions were made with some universal metrics (e.g., water loss indicators) that were considered too important to be excluded even if they were context dependant.

Performance indicators help AquaRating to establish a measure of current performance in the system for those aspects that can be quantified.

[2] Ocasionally, when the criterion to be assessed could not be measured, was heavily dependant on context or significant differences on how it was measured were found around the world, a good practice or a list of good practices was included to assess it.

Practices are simple yes/no questions that are more focused on the processes than on the outcome. As a result, practices cannot be considered a good measure of performance. However, good practices should lead to good performance, and therefore they may be taken as a proxy to assess future performance. Additionally, the quality of data needed to validate a practice is almost always far less demanding than a similar level of data quality for a variable.

[3] As part of the debate on the importance of adequate information, the team in charge of designing AquaRating decided to include data quality as the third pillar on which the rating rests. Therefore, the system goes beyond the recommendation of recording the quality of the data used to assess performance and in addition, it makes it an integral part of the assessment.

The assessment elements are organized into 8 main areas that constitute the key criteria used to determine the rating (see Figure 1). Each rating area is additionally subdivided into sub-areas (Figure 2) that include different assessment elements (either performance indicators or good practices). In practice, each element has a contribution to the rating of its area, and each area a contribution to the total rating. A system of weights determines the contribution of each individual element to the total rating.

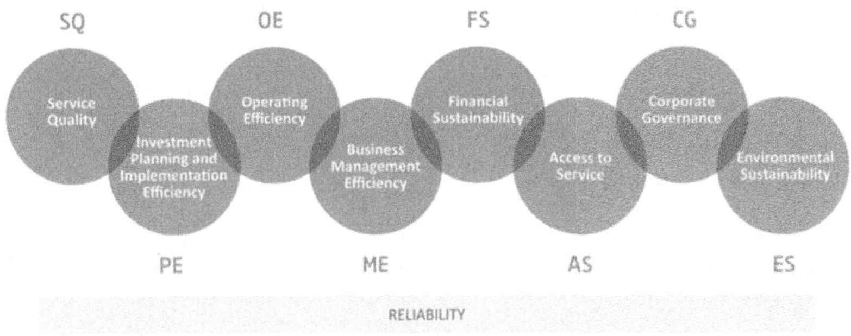

Figure 20. AquaRating rating areas.

Figure 21. Service quality hierarchy.

Adding the performance of good practices is simple, since they observe a digital behaviour (either yes or no) and therefore, a certain value is added (or not) to the rating. However, in the case of performance indicators, the value needs to be transformed into a value from 0 to 100 (making all weights comparable). This normalization is carried out by means of a function, which is particular for each performance indicator. Depending on the shape and values of this function (see Figure 3), the system can assess the different types of performance.

The potential contribution of a certain element to the rating still needs to be modified depending on the quality of the information used to produce the assessment. Different levels of data quality are specified for every single element of AquaRating. If the supporting documentation for a certain indicator corresponds to the highest possible level for data quality, the contribution of such indicator to the rating will only depend on the performance (and the normalization function). However, if the quality of the data feeding the variables corresponds to a lower level, a penalty will be applied (a figure from 0 to 1) to reduce the contribution of the indicator to the rating.

In other words, the maximum rating is possible by:

- Fulfilling all the good practices;
- Reaching the target value of 100 in the normalization function for each PI;
- Having the best possible quality of information for each assessment element.

Any deviations from the previous conditions will result in penalties that will reduce the "perfect" score.

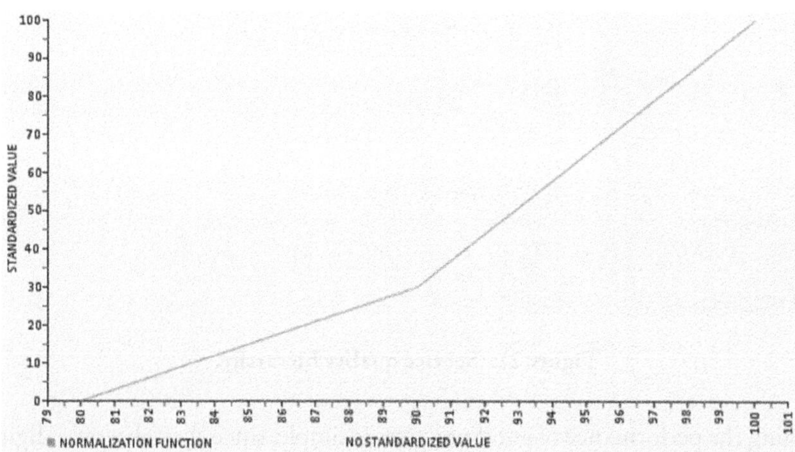

Figure 22. Standardization function for SQ1.5 compliance with drinking water standards.

In order to guarantee that the rating can be trusted, a third party must audit the data submitted to obtain an AquaRating. The audit takes place in several steps. Initially, the documents needed to validate the values and practices declared in AquaRating are uploaded into the web platform. A desk audit is then performed by the auditing firm, verifying the supplied information. In a second stage, an on-site audit is carried out, to make sure that the rating provided by the system is a correct one.

The whole AquaRating process is facilitated by the use of an online platform designed for the purpose. The software allows the management of users from an organization, calculation of the rating, upload of all the supporting documentation and communication between the auditors, the users and the AquaRating entity.

Summary of the performance assessment system

The AquaRating system has five levels:

- *Level 1*: Area
- *Level 2*: Sub-area
- *Level 3*: Assessment element (either a list of practices or an indicator)
- *Level 4*: Assessment sub-element (practices or variables)
- *Level 5*: Reliability

The following is a summary of the elements present in the AquaRating system:

- 8 areas, with 28 sub-areas
- 112 assessment elements, including 52 lists of good practices and 60 performance indicators
- 381 individual good practices
- 101 variables

Results and learnings

At the time of writing these lines, the only results that can be conveyed from AquaRating are those obtained during the pilot phase of the tool. However, the rigour and dimension of the pilot (with 12 services from several countries, differing in size and management type) equates to many performance assessment projects.

As with other projects, utilities found out that their corporate knowledge sometimes was lesser than expected. A lot of resources were used to gather the necessary information to feed the system and, often, auditable data was difficult to find.

Both the desk and on-site audits were quite new to almost everyone involved. The conclusions on them were really positive. Participating utilities highlighted the fact that the system could not be "cheated". The top management of some participating utilities appreciated the fact that their organizations had been pushed to produce a clear, accurate picture of their current status and the information from that picture was useful.

Finally, participants considered AquaRating not just an assessment tool, but also a learning one. The list of good practices provided a menu of potential improvements that was helpful in designing improvement plans for the services.

With the first ratings, AquaRating may become the first fully-audited assessment of urban water services carried out in a systematic way. This will contribute to better understand and have a greater knowledge of the sector.

8.5 Germany

(Author: Wolf Merkel, IWW Water Centre)

Country/Region	Germany
Scope	Drinking water
Objective	Global utility management, public and political domain (in parts)
Leading partner(s)	German Water Associations DVGW, BDEW, VKU; German Federal States
Participating utilities	Expected number of utilities (before 1st trial): up to 500
Character of the participation	Volunteer
Public availability of results	Partly
Data audits	Internal (within the decentralised projects)

Context and background

The German Water Supply sector started their benchmarking activities in 2001, based on the 1st edition of the IWA PI-system for water supply industry from 2000. A major milestone was the publication of the 1. German edition (Hirner & Merkel, 2005), resulting from a national research project with 13 water suppliers and implementation projects in German Federal States with more than 100 participants.

The German water sector is characterized by a multiplicity of operators with versatile operating models and - in the absence of a formal economic regulatory body – a high degree of self-organization. As a matter of fact, the adaptation and development of PI systems then took place in decentralised projects in the German federal states over the next 10 years. Progress and innovation has took place in terms of practical applicability and new meaningful indicators, with the drawback of certain deviations from the original PI system with regard to indicators, definitions and scope of application.

Against this background, the leading water associations DVGW, BDEW und VKU commissioned the IWW Water Centre in 2013 with the development of a three-tier **Water Sector Performance Indicator System (WS-PIS)**, consisting of **structural parameters**, **key performance indicators** and **key sector indicators** based on the IWA performance indicator system for water suppliers (Figure 23). This WS-PIS will be established as technical standard for *benchmarking* of water utilities and for *information and communication* in the political and public domain:

- Structural parameters in order to account for structural conditions in water supply (mainly context information for interpretation of benchmarking results and communication (Merkel *et al.* 2011);
- Relevant key sector indicators to monitor and communicate performance on reliability, quality, sustainability, customer service and economic efficiency of the water supply sector;
- Relevant key performance indicators for comparative performance assessment of water utilities.

Principal objective of the WS-PIS is the harmonization of benchmarking activities in the German Water Supply sector, providing a common denominator for individual project approaches, but allowing for regional adaptations and extensions.

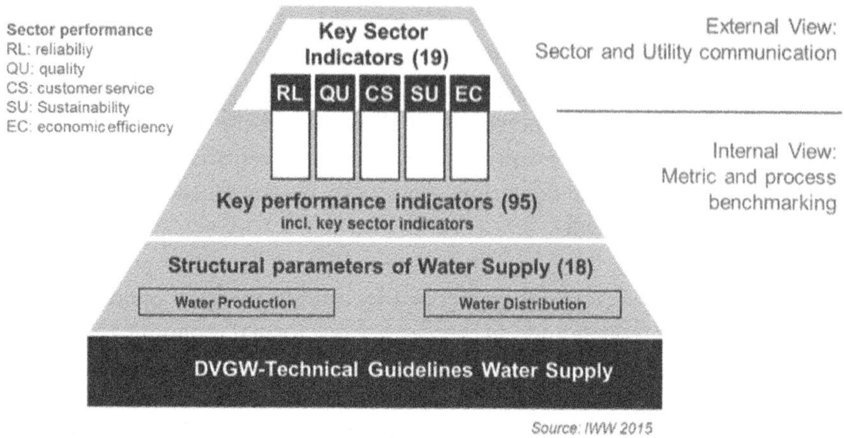

Figure 23. Structure of the harmonized German WS-PIS.

The main benefits for the water supply sector are regarded as follows:

- Harmonised performance indicators and data definition facilitate data collection and data quality assurance for the utilities
- Comparable results beyond individual projects allow benchmarking among utilities all over Germany
- Results from regional federal states' benchmarking projects can be easily consolidated for national sector reporting

Performance assessment process

Transparency and usage of the full range of expertise were the dominating features of the development process. From Oct 2013 to June 2014, all relevant stakeholders were invited to jointly develop the key indicators, resulting in a total number of 40 key experts from 26 organisations – companies, consultants, associations and authorities. The outcome of this joint development process was summarised by IWW in a public report, with in total 95 key performance indicators plus 18 structural parameters (details below). This report was then used as input for a set of technical guidelines namely DVGW W 1100-1 to W 1100-3, currently under public consultation. The scope of the PIS is utility benchmarking, while using the structural parameters for the clustering of utilities according to their individual context in water production and water distribution.

The WS-PIS will be piloted in several Federal States throughput the next 3 years. These Federal benchmarking exercises are commissioned by the State ministries to consultants with experience in this field, responsible for data acquisition with software/web support, the quality control of data, auditing, calculation and interpretation of the results. Reporting is usually two-split in a confidential company report and a public summary with aggregated results and findings.

Summary of the performance assessment system

The WS-PIS builds on the experience of a 10-years development process based on the IWA PI system, the 2nd edition published in 2006. The WS-PIS consists of 19 key sector indicators and 76 key performance indicators, plus a set of 18 structural parameters.

The key sector and key performance indicators are structured in five central performance dimensions of water supply, namely reliability, quality, customer service, sustainability and economic efficiency. Figure 24 shows the number of indicators attributed to each dimension.

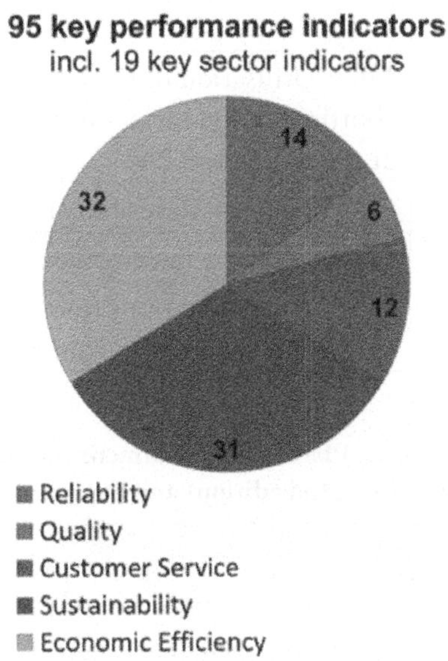

Figure 24. Indicators/context information attributed to the five key performance dimensions of water supply.

In technical terms of the IWA nomenclature the whole WS-PIS has 85 performance indicators and 28 context information. If compared to the set of the 2nd edition of IWA PIs (Alegre *et al.* 2006), the structural integrity of the system (water balance, tasks and management functions, definitions) remained unchanged, whereas a number of PIs have been added (mostly adding a detailed view or shifting to a more general view on a certain topic (Table 53, Table 54, Table 55, Table 56, Table 57).

Table 53. Overview on the PIs/CIs for the dimensions "reliability", indicating original IWA indicators (2nd edition) and new/modified indicators.

Reliability (14 PIs)	
IWA	WR2, Ph1, Op31, Op32, Op33, Op40
New/modified	WR5 Utilisation of legally secured water resources at the annual daily peak demand (%)
	WR8 Utilisation of legally secured annual own resources (%)
	WRN9 Utilisation of the annual imported water allowance (%)
	Ph3_1 Treated water storage capacity at the annual daily peak demand (%)
	Ph24 Utilisation of the overall supply system (bottleneck of abstraction/treatment/distribution) at the daily peak (%)
	Op51 Valves failures (no./1000 valves)
	QS13_1 Water interruptions (min)
	QS14_1 Interruptions (more than 3 hours) per connection (no./1000 connections)

Table 54. Overview on the PIs/CIs for the dimension "quality", indicating original IWA indicators (2nd edition) and new/modified indicators.

Quality (5 PIs + 1 CI)	
IWA	Op3, Op4, Op28, Op29, QS18
New/modified	CI Management systems (certified or accredited quality management, risk management schemes) (y/n)

Table 55. Overview on the PIs/CIs for the dimension "customer service", indicating original IWA indicators (2nd edition) and new/modified indicators.

Customer Service (8 PIs + 4 CIs)	
IWA	QS26, QS32
New/modified	QS36_1 Water supply resumed after connection failure within 12 hours (%)
	QS38 Customers allowing for direct debit authorization (%)
	Fi57_1 Customer service functions costs per customer (EUR/customer)
	QS42 Customer satisfaction with drinking water quality (grade)
	QS43 Customer satisfaction with service quality (grade)
	QS44 Customer satisfaction with value for money (grade)
	CI125 Securing standard supply standards (y/n)
	CI119 Customer complaints collection system (y/n)
	CI120 Customer complaints management system (y/n)
	CI Customer survey within the last 3 years (y/n)

The major alterations were observed i) in financial indicators, allowing for alternative accounting standards being used in public and private companies differently; ii) in alternative indicators for water losses and interruptions and iii) on streamlining with the existing German technical guidelines on infrastructure maintenance, water resources and energy efficiency.

Results and learnings

Pilot applications of the WS-PIS are anticipated in several Federal State's benchmarking activities within the next 3 years. The experience with the defined PIs and structural parameters will then be compiled and will be used for a further improvement of the whole system. With the harmonised WS-PIS, a higher degree of comparability between different water utilities is expected, enabling a more effective benchmarking process for the utilities and a more transparent look at the overall sector's performance.

Table 56. Overview on the PIs/CIs for the dimension "Sustainability", indicating original IWA indicators (2nd edition) and new/modified indicators.

Sustainability (27 PIs + 4 CIs)	
IWA	Pe19, Pe22, Pe24, Op16 (Op16* on 10-years average), Op20, Fi27, Fi46, Fi30 (Fi30* on expense basis), Fi33
New/modified	Pe31 Apprentices per total number of employees (%)
	Ph16 Water protection area per total catchment area (%)
	CI160 Water protection area with secured legal status (%)
	CI161 Water protection area with legally binding protection contracts (on farming etc.) (%)
	CI-WR10 Raw water quality index (–)
	CI-WR11 Raw water availability index (–)
	Ph25 Total energy consumption (kWh/m^3)
	Ph5_1 Standardised energy consumption (per 100 m geodetical height not pump head) (kWh/m^3/100 m)
	Ph17_1 Energy consumption for water pumping (kWh/m^3) (alternative to Ph5_1)
	Ph21 Energy consumption water treatment (%)
	Ph22 Energy consumption water distribution (%)
	Ph23 Energy consumption from renewable resources (incl. own recovery) (%)
	Op55 Sustainable net rehabilitation index (rehab rate/loss index/failure index) (–)
	Op56 Mains rehabilitation-renovation-replacement (%/year)
	Op57 Mains rehabilitation-renovation-replacement (10-years average) (%/year)
	Fi60 Water resources and catchment management costs per water abstracted (€/m^3) (Fi60* on expense basis)
	Fi79 Sustainable water management costs (investments and running costs) (€/m^3)
	Fi81 Average yearly unit investment water production per main lengths (10-years average) (€/m^3)
	Fi82 Average yearly unit investment water distribution per treated water volume (10-years average) (€/km)
	Fi85 Equity ratio (–)

Table 57. Overview on the PIs/CIs for the dimension "economic efficiency", indicating original IWA indicators (2nd edition) and new/modified indicators.

Economic efficiency (32 PIs + 1 CI)	
IWA	Pe2, Pe7, Fi1 (+Fi1* earnings), Fi4 (+Fi4*), Fi5 (+Fi5*), Fi6 (+Fi6*), Fi7, Fi8, Fi9, Fi10 (electric energy only), Fi23, Fi24, Fi16, Fi11
New/modified	Pe27 Administration personnel (%)
	Fi73 Variable sales revenues (%)
	Fi74 Invariable sales revenues (%)
	Fi85a Specific total expenses of water supply: normalized by external factors (fees etc.) (€/m^3)
	Fi85b Specific total costs water of supply (normalized by external factors) (€/m^3)
	Fi48_1 Material costs (€/m^3)
	Fi50_1 Water abstraction fees (€/m^3)
	Fi51 Concession fees (€/m^3)
	Fi54 Administration costs (€/m^3)
	Fi78 Water production costs (€/m^3)
	Fi80 Water distribution costs (€/km)
	Fi87 Water tariffs as average annual water costs per customer (€/person)
	CI-Fi86 Standardized accounting system (y/n)

Case 4 references

Alegre H., Baptista J. M., Cabrera E. Jr., Cubillo F., Duarte P., Hirner W. et al. (2006). Performance Indicators for Water Supply Services. IWA (ed.), IWA Publishing, London, p. 289.

ATT, BDEW, DBVW, DVGW, DWA, VKU. (2015). Profile of the German Water Sector: Summary. Weblink (10/2015). http://www.vku.de/index.php?eID=tx_nawsecuredl&u=0&g=0&t=1443759886&hash=e22d3f888c222d3b13aadd4e608db596b1dcb97b&file=fileadmin/media/Dokumente/Oeffentlichkeitsarbeit_Presse/Publikationen/2015/WEB_brachenbild_ENGL_wasserwirtschaft_2015_a4_KURZFASSUNG_16062015.pdf

Hirner W. and Merkel W. (2005). Kennzahlen für Benchmarking in der Wasserversorgung. WVGW Wirtschafts- und Verlagsgesellschaft Gas und Wasser mbH, Bonn, ISBN/ISSN 3-89554-152-4, 348 S.

Merkel W., Staben N., Bräcker J., Levai P., Neskovic M., Petry D. and Weiss M. (2011). Structural conditions in German water supply. Water Science and Technology: Water Supply. **11**(5), 596–604.

8.6 Case 5: TRUST/AWARE-P: from incipient to leading-edge IAM planning implementations

(Author: Helena Alegre, LNEC, Portugal)

This case shows how the IWA PI systems were instrumental for the development of the AWARE-P IAM approach and software (www.aware-p.org), and subsequent utility implementation and use (www.iniciativaGPI.org). It also refers to the work developed in the European Research project TRUST (www.trust-i.net).

Country/region	• Europe
Scope	• Urban water services
Objective	• Global utility management
Leading partner(s)	• AWARE-P coordinator: LNEC (Portugal) • iGPI coordinator: LNEC (Portugal) • TRUST coordinator: IWW (Germany)
Participating utilities	• Full implementation in 30 utilities in Portugal • Exploratory cases in the USA, Spain, Australia, Angola and Brazil
Character of the participation	• Volunteer. Note: In Portugal, utilities serving more than 30000 people are legally required to implement an IAM system.
Public availability of results	• Not in detail. Papers available with global and selected results.
Data audits	• Internal, supported and complemented by LNEC

Context and background

Learning from leading know-how and practice (e.g., Australia, New Zealand, EU research and IWA's manuals of best practice and network of professionals),

LNEC partnered with other R&D organizations, software developers and utilities in the AWARE-P project to jointly develop a ground-breaking IAM planning methodology, supported by technical guides and a professional-grade, innovative software, plus e-learning and demo cases.

The R&D stage was followed by a rollout & capacitation stage (2012-), where each utility develop internally, approve and implement their strategic and tactical IAM plans using the AWARE-P methodology, software and training materials.

Table 58. The water & wastewater utilities in the rollout stage.

2012–2013: 30 utilities (IAM plans concluded and being implemented)
Ág. Coimbra, AdRA, AQUAPOR – Ág. Planalto, CM Sabugal, EMAR Vila Real, INDAQUA, Infraquinta, Inframoura, Infralobo, INOVA, SM Abrantes, SMAS Almada, SM Castelo Branco, SM Loures, SMAS Sintra, SMSB Viana do Castelo, Águas do Oeste, AGERE, EAmb Esposende Águas de Alenquer, Águas de Barcelos, Águas de Cascais, Águas da Covilhã, Águas da Figueira, Águas de Gondomar, Águas do Marco, Águas do Sado, Águas da Serra, AGS Paços de Ferreira, Tratave.
2014–2015: 30 utilities, including 10 new and the others from the previous projects (advanced mode).
New utilities: AGERE, Águas do Porto, Águas de Santarém, Águas do Zêzere e Côa, Aguas de Valencia, AQUALIA – Cartagua, CM Barreiro, CM Palmela, SMAS Vila Franca de Xira, SANEST.

Today, over 25% of Portugal's population are already served by utilities that have produced their corporate strategic and tactical IAM plans based on the AWARE-P methodology and templates — endorsed by the national water services regulator ERSAR — with widespread use of the open-source software, effectively creating alignment and critical mass for much-needed change. This has been supplemented by 2 utilities in Spain and one in the USA, with a pilot in Australia; while the AWARE-P software has over 1000 registered users in 5 continents. IWA's LESAM 2013 singled out Portugal's disruptive IAM program in its conclusions, while WERF (USA) is promoting uptake of the innovative methodology and further development of the software in the USA, and the World Bank's Danube Water Program has expressed an interest in the rollout formula utilized in the project. Given the diversity and representativeness of Portugal's water market, this project provides a valuable blueprint for similar development elsewhere, as recognized by the 2014 Project Innovation Award 2014 (Planning; Europe and West Asia Region) and Mulheim Water Award 2014.

The roll-out stage runs in parallel with new research. The European research project TRUST – Transition for the urban water services of tomorrow, is the most important instance. In TRUST, the concept of sustainability was revisited and key sustainability objectives and criteria were established (www.trust-i.net>downloads). The IWA PIs were rearranged according to these objectives and criteria, being now part of the TRUST/AWARE-P PI portfolio.

Sharing the same roots and drivers, the IWA IAM approach and the TRUST/AWARE-P approach coincide in principles and rationale. Performance assessment plays a fundamental role in the whole process. It is used to define targets for each objective, to carry out a diagnosis, to prioritize areas of intervention, to compare and select alternatives of intervention and to monitor results. The core of the performance metrics adopted by the utilities referred above is composed of IWA PIs.

Performance assessment process

The first stage of IAM plan developments by the utilities is the definition of clear objectives and assessment criteria, endorsed by the top managers. The selection of metrics follows. The utilities are naturally free to use their own indicators, but the IWA PI system has proven to be their main reference. The Portuguese Regulatory system (case 1) adopts mostly IWA PIs. Utilities tend to include these PIs in their own management systems. Each metric is clearly defined and data collection procedures are documented; a small number of metrics is recommended; most utilities adopt between 10 and 30 metrics. Reference values are established.

A diagnosis is carried out based on this assessment system for the whole system, as well as for its composing functional units (areas of analysis); Targets are established for each metric. Areas of analysis are ranked in terms of priority of intervention.

Every high priority area is analysed in more detail, alternatives on intervention are designed, compared and selected, using again the same (or an adjustment version when appropriate) performance assessment system.

IAM plans are prepared, approved and implemented.

Monitoring of plan implementation is again carried out based of the same performance assessment system.

The AWARE-P open-source software

Utilities are free to use any software (Coelho *et al.* 2013), but the methodology and the analysis methods developed are supported by the AWARE-P IAM planning software (published by baseform.org). AWARE-P is an open environment for infrastructure asset management planning, supporting the quantifiable impact assessment of AM interventions and promoting the best trade-offs between performance, economics and risk. AWARE-P offers the PI tool, which contains libraries with the IWA PIs and variables.

Results and learnings

The project's products included guidelines for developing IAM strategic and tactical plans, including MS Word® templates; training materials (presentations slides, recorded webinars, written materials); significant improvements in the AWARE-P software; a national visibility public event where the participants reported mid-term results (in March 2013; around 200 attendants); papers in industry publications (e.g., Alegre *et al.* 2013; Leitão *et al.* 2014; Cardoso *et al.* 2015).

Central among the project results were the strategic and tactical IAM plans that each of the utilities produced, an impressive batch creating enough precedent for a broader roll-out (Leitão *et al.* 2015; Cardoso *et al.* 2015). The way in which the several participating utilities developed their strategic and tactical plans, and implemented their IAM processes, was shaped by their context, resources, organizational maturity, information availability and the existence of other management systems or instruments. A large variety of cases illustrating the diversity of situations and solutions are available, two of which are illustrated in Table C1.3.

For the participating utilities, the project has led to effective harmonization of the organization's objectives, among the various management systems in place, with considerable alignment between decision levels and management processes (e.g., harmonizing kPIs cross-levels); the IAM strategic plans developed under iGPI are used as a central management tool in several of the participants; the project and IAM processes have been powerful drivers for improvements in data depth, quality, focus and integration; and many enhanced procedures for data collection, particularly related to work orders and surveying of buried assets, have been implemented.

Figure 25. Strategic IAM template and plans produced.

The project has also had a major impact on the water industry, in Portugal and internationally, helping to raise awareness of the issue and making available software tools, documentation, planning templates, and a networkable community of IAM practitioners. The project provides the water services regulator with a workable basis to further develop the current regulatory framework and public policies.

The variety of cases covered demonstrates how, in actual practice, systematic and well-devised IAM processes can be tailored and successfully implemented in many different contexts. The collaborative project format has proven to be particularly suited to the task, enabling a quicker and more effective cultural change, technical uptake and process implementation.

The maturity of the Portuguese water industry's PI culture, grounded on over 15 years' experience with the IWA PI-Field test and the National regulatory PI system, has been a major stepping stone and a critical success factor for the incorporation in the utilities of the IWA IAM approach.

Table 59. The iGPI program: objectives and outcomes.

<table>
<tr><td>CASE 1
Mid-size utility</td><td>Features: technologically developed utility; well trained staff; good inventory; complete, reliable GIS; good monitoring systems; hydraulic models available for the entire water supply system.</td></tr>
<tr><td colspan="2">• The availability of a large amount of information, mostly reliable, including all-mains calibrated hydraulic models for the entire water supply system, allowed this utility to use more sophisticated and data-demanding metrics to address aspects such as pressure adequacy and flow velocity adequacy.</td></tr>
<tr><td colspan="2">• Automated procedures have been implemented in order to calculate the selected metrics.</td></tr>
<tr><td colspan="2">• Some metrics at the strategic level result from the aggregation of more detailed metrics adopted at the tactical level.</td></tr>
<tr><td colspan="2">• Despite the technological maturity, the use of work orders information for reliability analysis revealed room for improvement. Non-infrastructural tactics were established to address this problem.</td></tr>
<tr><td colspan="2">• With an IAM metrics system in place, the fact that a significant part of the process has been automated shortens the time and manpower needed for detail diagnosis, which allow this utility to work simultaneously with 4 pilot network sectors at the tactical level (2 each for water supply and for wastewater) during the course of the project.</td></tr>
<tr><td>CASE 2
Large utility</td><td>Features: the utility is the result of the recent merger of 10 municipal water and wastewater services; very diverse contexts, challenges, and data availability and quality among the municipalities; certified BSC management system; GIS reflects the municipalities' disparity in data availability, depth and quality.</td></tr>
<tr><td colspan="2">• iGPI was seen by the utility as a golden opportunity to help establish sound organizational processes.</td></tr>
<tr><td colspan="2">• One of the challenges for this utility was to prioritize the municipalities with higher rehabilitation needs, in a sound, transparent and accountable way, in a context of local political sensitiveness where consensus-driven negotiation is crucial. The results from iGPI helped respond to this challenge, particularly through a sound IAM metrics system.</td></tr>
<tr><td colspan="2">• The existing BSC implementation did not address the long-term effect of the managerial decisions. Although changes are more difficult to implement given the size and complexity of the organization, several new metrics have been included and a transition process has been devised towards a fully satisfactory BSC implementation.</td></tr>
<tr><td colspan="2">• iGPI also gave rise to multiple new data collection procedures, particularly related to GIS, work orders, and harmonization among information systems.</td></tr>
</table>

Case 5 References

Alegre H., Brattebø H., Cabrera E. and Hein A. (2012). Framework for sustainability assessment of UWCS and development of a self-assessment tool. TRUST – Transitions to the Urban Water Services of Tomorrow, final deliverable D31.1, http://www.trust-i.net/downloads/index.php?iddesc=48.

Alegre H., Coelho S. T., Covas D., Almeida M. C. and Cardoso M. A. (2013). A utility-tailored methodology for integrated asset management of urban water infrastructure. Water Science & Technology: Water Supply, **13**(6), 1444–1451 © IWA Publishing 2013. Doi:10.2166/ws.2013.108.

Cardoso M. A., Silva M. S., Poças A., Ribeiro R., Almeida M. C., Coelho S. T. and Alegre H. (2015). Innovation results of IAM planning in urban water service. LESAM 2015, IWA, Yokohama, Japan.

Coelho S. T., Vitorino D. and Alegre H. (2013). AWARE-P: a system-based software for urban water IAM planning. IWA LESAM 2013, 10–12 Sep, Sydney, Australia.

Leitão J. P., Coelho S. T., Alegre H., Cardoso M. A., Silva M. S., Ramalho P., Ribeiro R., Covas D., Vitorino D., Almeida M. C. and Carriço N. (2014). Moving urban water infrastructure asset management from science into practice. Urban Water Journal, DOI: 10.1080/1573062X.2014.939092.

9. REFERENCES AND SELECTED BIBLIOGRAPHY

Alegre H. (2002). "Performance indicators as a management support tool". In Urban Water Supply Handbook, L. W. Mays (ed.), McGraw-Hill, ISBN 0-07-137160-5.

Alegre H. and Almeida M. C. (2009). Strategic Asset Management of Water and Wastewater Infrastructures. IWA Publishing, ISBN 97843391869.

Alegre H. and Baptista J. M. (ed.) (1997). Workshop Performance Indicators for Transmission and Distribution Systems, International Water Association, Lisbon, Portugal.

Alegre H., Baptista J. M. and Coelho S. T. (2005). "Performance Indicators for water network rehabilitation". In CARE-W – Computer Aided Rehabilitation for Water Networks, S. Sægrov (ed.), EU project: EVK1-CT-2000-00053, ISBN: 1843390914, IWA Publishing.

Alegre H., Hirner W., Baptista J. M. and Parena R. (2000). Performance Indicators for Water Supply Services. Manual of Best Practice Series, IWA Publishing, London, ISBN 1 900222 27 2.

Brown R. E. and Humphrey B. G. (2005). Asset Management for Transmission and Distribution. Power and Energy Magazine, IEEE, **3**(3), 39.

Cabrera E. (2001). Diseño de un sistema para la evaluación de la gestión de abastecimientos urbanos. PhD Thesis, Polytechnic University of Valencia, Spain.

Cabrera E. Jr., Dane P., Haskins S. and Theuretzbacher-Fritz H. (2011). Benchmarking Water Services. IWA Publishing, London, ISBN: 1843391988

Duarte P., Alegre H. and Matos R. (2003). About the two IWA performance indicators systems for urban water management. In PEDS 2003 – Pumps, Electromechanical Devices and Systems Applied to Urban Water Management Conference Proceedings, Valencia, Spain.

Farley M. and Troe S. (2003). Losses in water distribution networks. A Practitioner's Guide to Assessment, Monitoring and Control. IWA Publishing, ISBN: 1-900222-11-6.

Guérin-Schneider (2001). Introduire la mesure de performance dans la régulation des services d'eau et d'assainissement en France – Instrumentation et organisation. PhD Thesis, Laboratoire GEA, ENGREF. Paris, France.

IAS/IFRS (International Accounting Standards/International Financial Reporting Standards enacted by the International Accounting Standard Board – IASB).

ISO 24510 (2007). Activities relating to drinking water and wastewater services – Guidelines for the assessment and for the improvement of the service to users.

ISO 24511 (2007). Activities relating to drinking water and wastewater services – Guidelines for the management of wastewater utilities and for the assessment of drinking water services.

ISO 24512 (2007). Service activities relating to drinking water and wastewater – Guidelines for the management of drinking water utilities and for the assessment of drinking water services.

IWA (2000). Blue Pages on Losses from Water Supply Systems. A. Lambert e W. Hirner (ed.), International Water Association (edição electrónica).

Lambert A., Myers S. and Trow S. (1998). Managing Water Leakage – Economic and Technical Issues. Financial Times Energy, London.

Lobato de Faria A. and Alegre H. (1996). Paving the way to excellence in water supply systems: a national framework for levels-of-service assessment based on consumer satisfaction, The Maarten Schalekamp Award - 1995, AQUA, **45**(1), 1–12. February 1996, IWSA, London, UK.

Matos R., Cardoso A., Ashley R., Duarte P., Molinari A. and Schulz A. (2003). Performance Indicators for Wastewater Services, IWA Publishing, ISBN: 1900222906, p. 192.

Merkel W. (2002). "International Report: Performance assessment in the water industry". In Water Science and Technology: Water Supply, **2**(4), 151–162, IWA Publishing, London.

Neely A., Adams C. and Kennerley M. (2002). Performance Prism: The Scorecard for Measuring and Managing Stakeholder Relationships. Financial Times Prentice Hall.

OFWAT (2006). Levels of Service for the Water Industry in England & Wales 2004-2005 Report. Retrieved March 2006, from the Office of Water Services (OFWAT) Web site: www.ofwat.gov.uk.

Parena R., Smeets E. and Troquet I. (2002). Process Benchmarking in the Water Industry. Manual of Best Practice Series, IWA Publishing, London, ISBN 9781843390107.

Ramsey S. (2003). The expansion of the world bank's international benchmarking network for water and sanitation utilities. Proc. Conference "Global Developments in Water Industry Performance Benchmarking", 29 Sept.–2 Oct. 2003, Perth, Austrália.

Stahre P. and Adamsson J. (2003). Performance Benchmarking: A Powerful Management Tool For Water and Wastewater Utilities. Presented at AFNOR Atelier « Assurer la qualité de la gestion des services d'eau potable et d'assainissement: les bonnes pratiques – un enjeu, une méthodologie, un partage d'expériences », Paris, France.

Thorton J. (2002). Water Loss Control Manual, McGrawHill, ISBN 0-07-137434-5.

UKWIR Ltd (2002). Capital Maintenance Planning: A Common Framework. Vol. 1 – Overview. 02/RG/05/3 ISBN 1-840572655.

PART II – SPECIFICATION OF THE IWA PI SYSTEM

1. DEFINITIONS

1.1 Water balance definitions

Definition of water supply system inputs and outputs

A well-established water balance is fundamental for water losses assessment. The definitions, terminology and choices of performance indicators for water losses used in this document are based substantially on the work of the IWA Task Force on "Water losses", with some minor adaptations to allow them to be fitted within the overall context of this wider performance indicators report. The work of the Task Force on "Water losses" is described more fully in an AQUA paper (December 1999), an IWA Leaflet (December 1999) and a Blue Pages on Losses from Water Supply Systems: "Standard Terminology and Performance Measures" (2000).

Figure 26 illustrates the principal inputs and outputs to a typical water supply system sequentially, from raw water intake to consumption by customers. Some systems will of course be simpler and will not have all of the features shown.

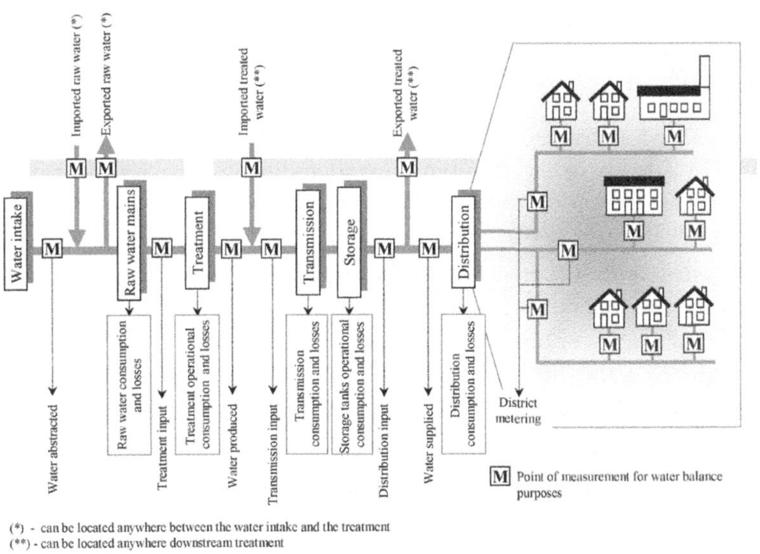

Figure 26. Definition of water supply system inputs and outputs.

The water balance requires estimates of volumes of water to be made at each measurement point applicable to the system under consideration. Where there are actual meters, data from these would normally be used, but in the absence of meters, a 'best estimate' based on other related available data and application of sound engineering judgement may be required.

The water balance is normally computed over a 12 month period and thus represents the annual average of all components.

Definitions relating to Figure 26, and Table 60 which follows in the next Section, are given below. Because of widely varying interpretations of the term Unaccounted-for Water (UFW) worldwide, the Task Force on "Water losses" and the Task Force on "Performance indicators" do not recommend use of this term. If the term is used at all, it should be defined and calculated in the same way as *non-revenue water* in Table 60.

Water abstracted: the volume of water obtained for input to water treatment plants (or directly to the transmission and distribution systems) that was abstracted from raw water sources during the assessment period.

Raw water, imported or exported: the volumes of bulk transfers of raw water across operational boundaries during the assessment period. The transfers can occur anywhere between the abstraction point and the treatment plants.

Treatment input: the volume of raw water input to treatment works during the assessment period.

Water produced: the volume of water treated for input to water transmission lines or directly to the distribution system during the assessment period. The volume of water that is distributed to consumers without previous treatment shall be also accounted for in *water produced*.

Treated water, imported or exported: the volumes of bulk transfers of treated water across operational boundaries during the assessment period. The transfers can occur anywhere downstream treatment.

The volume of water (if any) that is abstracted and delivered to consumers without any treatment shall also be accounted for as treated water in the scope of the water balance.

Transmission input: the volume of treated water input to a transmission system during the assessment period.

Distribution input: the volume of treated water input to a distribution system during the assessment period.

Supplied water: the *distribution input* minus *treated water exported*. (When it is not possible to separate transmission from distribution, supplied water is the *transmission input* minus *treated water exported*).

System input volume: the volume input to water supply system during the assessment period.

Note: If the water balance calculation relates to a part of the global system, the system input volume should correspond to that part.

Authorised consumption: the volume of metered and/or non-metered water taken by registered customers, the water supplier and others who are implicitly or explicitly authorised to do so by the water supplier, for residential, commercial and industrial purposes, during the assessment period. It includes *water exported.*

Note (1): Authorised consumption may include items such as fire fighting and training, flushing of mains and sewers, street cleaning, watering of municipal gardens, public fountains, frost protection, building water, etc. These may be billed or unbilled, metered or unmetered, according to local practice.

Note (2): Authorised consumption includes leakage and waste by registered customers that are unmetered.

Real losses: physical water losses from the pressurised system, up to the point of measurement of customer use during the assessment period. The volume lost through all types of leaks, bursts and overflows depends on frequencies, flow rates, and average duration of individual leaks.

Note: Although physical losses after the point of customer flow measurement are excluded from the assessment of real losses, they are often significant (particularly where customers are unmetered) and worthy of attention for demand management purposes.

Apparent losses: accounts for all types of inaccuracies associated with production metering and customer metering, plus unauthorised consumption (theft or illegal use).

Note: Under-registration of production meters, and over-registration of customer meters, leads to under-estimation of real losses. Over-registration of production meters, and under-registration of customer meters, leads to over-estimation of real losses.

Non-revenue water: the difference between the volumes of system input and billed authorised consumption. Non-revenue water includes not only the real losses and apparent losses, but also the unbilled authorised consumption.

Water balance components

Table 60 shows the recommended standard format and terminology for Water Balance calculations for one or more sections of a water supply system (e.g., raw water mains, transmission, and distribution). If original water balance data is available in any alternative format or terminology, it must be re-assembled into the components in Table 60, in volume per year, before attempting to calculate any performance indicators. The calculation associated with the Water Balance that is listed below Table 60 is shown in detail in section II-3.1 – Section A – Water volume data.

Table 60. Components of water balance.

<table>
<tr><th>A</th><th>B</th><th>C</th><th>D</th><th>E</th></tr>
<tr><td rowspan="10">System input volume

[m³/year]</td><td rowspan="4">Authorised consumption

[m³/year]</td><td rowspan="2">Billed authorised consumption

[m³/year]</td><td>Billed metered consumption (including water exported) [m³/year]</td><td rowspan="2">Revenue water

[m³/year]</td></tr>
<tr><td>Billed unmetered consumption [m³/year]</td></tr>
<tr><td rowspan="2">Unbilled authorised consumption

[m³/year]</td><td>Unbilled metered consumption [m³/year]</td><td rowspan="8">Non-revenue water

[m³/year]</td></tr>
<tr><td>Unbilled unmetered consumption [m³/year]</td></tr>
<tr><td rowspan="6">Water losses

[m³/year]</td><td rowspan="2">Apparent losses

[m³/year]</td><td>Unauthorised consumption [m³/year]</td></tr>
<tr><td>Metering inaccuracies water losses [m³/year]</td></tr>
<tr><td rowspan="4">Real losses

[m³/year]</td><td>Real losses on raw water mains and at the treatment works (if applicable) [m³/year]</td></tr>
<tr><td>Leakage on transmission and/or distribution mains [m³/year]</td></tr>
<tr><td>Leakage and overflows at transmission and/or distribution storage tanks [m³/year]</td></tr>
<tr><td>Leakage on service connections up to the measurement point [m³/year]</td></tr>
<tr><td colspan="5">Note: Consumption of water by registered customers who pay indirectly through local or national taxation is deemed to be billed authorised consumption for the purposes of the water balance.</td></tr>
</table>

Steps for calculating non-revenue water and water losses:

Step 1: Define *system input volume* and enter in Column A.

Step 2: Define *billed metered consumption* and *billed unmetered consumption* in Column D; enter total in *billed authorised consumption* (Column C) and *revenue water* (Column E).

Step 3: Calculate the volume of *non-revenue water* (Column E) as *system input volume* (Column A) minus *revenue water* (Column E).

Step 4: Define unbilled metered consumption and unbilled unmetered consumption in Column D; transfer total to unbilled authorised consumption Column C.

Step 5: Add volumes of billed authorised consumption and unbilled authorised consumption in Column C; enter sum as authorised consumption (Column B).

Step 6: Calculate water losses (Column B) as the difference between system input volume (Column A) and authorised consumption (Column B).

Step 7: Assess components of unauthorised consumption and metering inaccuracies (Column D) by best means available, add these and enter sum in apparent losses (Column C).

Step 8: Calculate real losses (Column C) as water losses (Column B) minus apparent losses (Column C).

Step 9: Assess components of real losses (Column D) by best means available (night flow analysis, burst frequency/flow rate/duration calculations, modelling, etc.), add these and crosscheck with volume of real losses in Column C.

Particular difficulty is experienced in completing the water balance with reasonable accuracy where a significant proportion of customers are not metered. Authorised consumption in such cases should be derived from sample metering of sufficient numbers of statistically representative individual connections of various categories and sub-categories and/or by measurement of total flows into discrete areas of uniform customer profile, also of various categories and sub-categories. In the latter method subtraction of leakage demand from total input is necessary, leakage being determined by analysis of the sub-components of night demand, adjusting for diurnal pressure variation as appropriate. The confidence grade allocated to the authorised consumption (see I-6 & II-5) should reflect the rigour of the investigations.

In this document it is the intention that performance related to management of water losses should be measured from three different viewpoints: financial, technical and water resources. If the water balance calculations proceed no further than Step 3 in Table 60, as is the case in the simplest traditional water balances, the only performance indicator that can be calculated is the financial one of:

Fi46: Non-Revenue Water by Volume

The Task Force on "Water losses" emphasised the importance of completing the calculation up to Step 8 (and preferably Step 9) in Table 60 – in particular, the importance of attempting, by the best means available, the separation of water losses into *apparent* and *real* losses. This allows the calculation of a required range of water resources, operational and financial performance indicators, as shown in Table 61:

Table 61. Code, recommended units and comments on the performance indicators for water losses and non-revenue water.

Indicator	Comments
Water resources water losses PI	
WR1 – Inefficiency of use of water resources (%)	Environmental indicator, unsuitable for assessing efficiency of the technical management of distribution systems.
Operational water losses PI	
Op23 – Water losses (m^3/service connection/year)	As it accounts for both apparent losses and real losses, sometimes it is more convenient to express total losses as % or as l/service connection/day.
Op25 & Op26 – Apparent losses (%)	For bulk supply systems, the indicator is expressed as a percentage of the total system input volume. For distribution systems, it is expressed as a percentage of the system input volume minus the exported water.
Op27 – Real losses (litres/service connection/day when system is pressurised)	Allows for intermittent supply situations.

Table 61. Code, recommended units and comments on the performance indicators for water losses and non-revenue water (Contd.).

Indicator	Comments
Op29 – Infrastructure leakage index (-)	Technical achievable low-level real losses are equal to the best estimate of so called Unavoidable Average Real Losses, UARL. They include system-specific allowance for connection density, customer meter location on service, and current average pressure. This indicator does not fulfil some of the requirements for a PI listed in 2.2.
Economic and financial water losses PI	
Non-revenue water by volume	Can be calculated from simple water balance.
Non-revenue water by cost	Allows for separate values/m^3 for components of non-revenue water.

1.2 Organisation functions

This section defines the organisation functions included in:

- general management;
- human resources management;
- financial and commercial;
- customer service;
- *technical services*:
- planning and construction;
- operations and maintenance;

that are used in several sub-sections of chapter II-2 — PERFORMANCE INDICATORS.

The following tables aim to support the definition of the personnel and financial indicators related to the organisation functions and refer exclusively to the core-business activities.

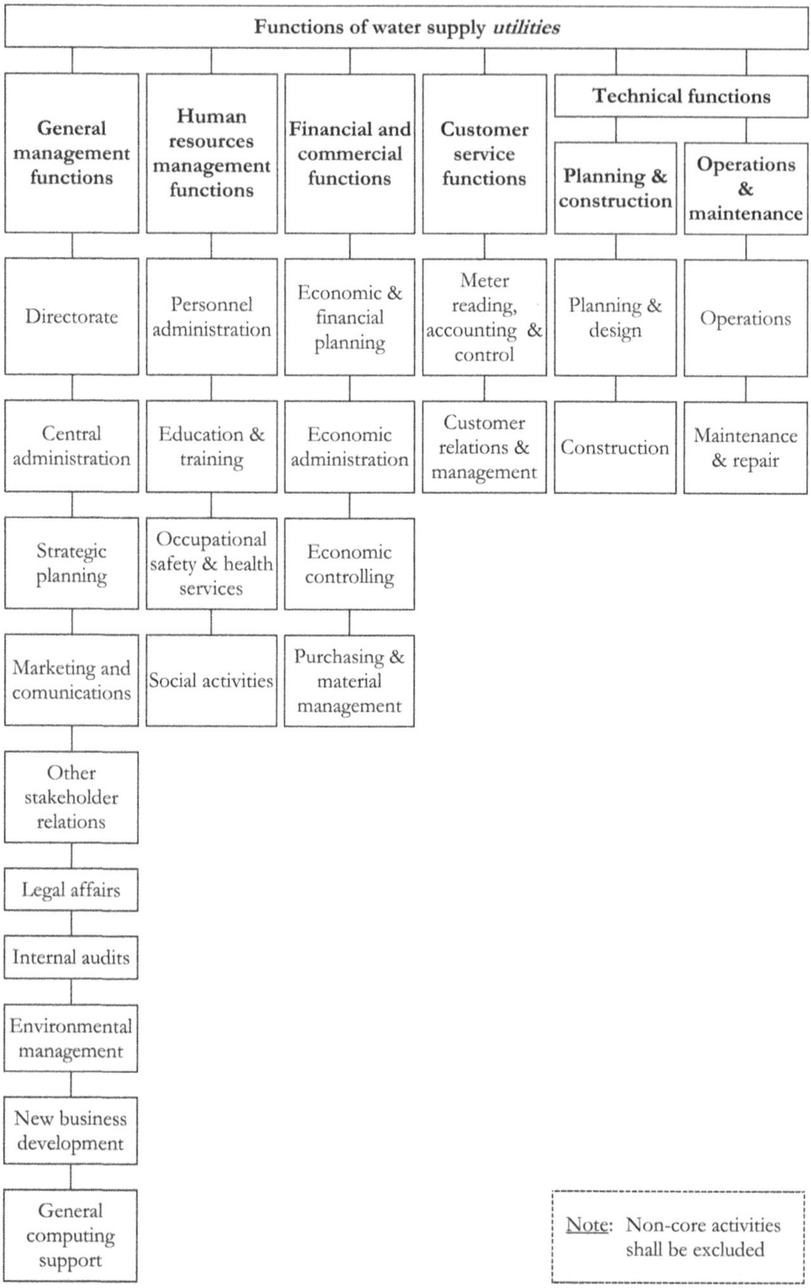

Figure 27. Functions of water supply *utilities*.

Tables below specify the sub-functions of each partial function indicated in Figure 27.

Table 62. Organisation functions – General management.

Main Functions	Partial Functions	Sub Functions
General management functions	Directorate	❑ Strategic policies ❑ External representation and relationships ❑ New business[7]
	Central administration	❑ General secretary ❑ Archiving
	Strategic planning	❑ Objectives definition ❑ Organisation ❑ Information systems planning ❑ Performance assessment
	Marketing and comunications	❑ Press service and releases ❑ Communication and image ❑ Customer surveys ❑ Public awareness campaigns ❑ *marketing* ❑ Publications and information ❑ Company reports
	Other stakeholder relations	❑ Shareholder affairs ❑ Policy-making bodies affairs ❑ Regulatory affairs ❑ Other stakeholders affairs
	Legal affairs	❑ Legal contracts ❑ Tenders documentation ❑ Legal protection and insurance ❑ Land administration ❑ Contentious matters and litigation

Table 62. **Organisation functions – General management (contd.).**

Main Functions	Partial Functions	Sub Functions
	Internal audits	❑ Check of financial procedures ❑ Check of administration procedures ❑ Check of technical procedures ❑ Quality controls
	Environmental management	❑ Environmental policies ❑ Environmental management and audits ❑ Control of waste management
	Business development[7]	❑ Expansion of services provided ❑ Expansion of existing customer base
	General computing support	❑ Management and operation of central computing resources and internal networks ❑ General management of the decentralised computing resources ❑ Software management, including intranet and internet use ❑ Design, implementation and management of information systems, excluding specific applications for technical functions ❑ Training and technical support to the *utility* users

[7] Cost and manpower for new business activities can be included while under development. All non-core business already in-place shall be excluded.

Table 63. Organisation functions – Human resources management.

Main Functions	Partial Functions	Sub Functions
Human resources management functions	Personnel administration	❑ Personnel planning and development ❑ Recruitment ❑ Personnel management ❑ Job evaluation ❑ Payroll accounting and payment
	Education and training	❑ Training planning ❑ Implementation and control of training ❑ Dissemination of technical information
	Occupational safety and health services	❑ Medical checks ❑ Evaluation of unfitness for work ❑ Working safety policies, implementation and control
	Social activities	❑ Integration of new personnel ❑ Social assistance ❑ Socio-cultural activities

Table 64. Organisation functions – Financial and commercial.

Main Functions	Partial Functions	Sub Functions
Financial and commercial functions	Economic and financial planning	❑ Investment planning and control ❑ Cost planning and control ❑ Revenue and cost recovery planning ❑ Financial planning and policies
	Financial administration	❑ Book keeping ❑ Water service assets assessment ❑ Annual accounts reporting
	Economic control	❑ Planning control ❑ Accounting control ❑ Economic performance assessment
	Purchasing and material management	❑ Suppliers information management ❑ Advertisement ❑ Commissioning ❑ Warehouse management

Table 65. Organisation functions – Customer service.

Main Functions	Partial Functions	Sub Functions
Customer service functions	Meter reading, billing and control	❑ Meter reading ❑ Billing ❑ Collection control ❑ Consumption control
	Customer relations and management	❑ Incoming order management ❑ Customer consultation ❑ Complaint management ❑ Information dissemination

Table 66. Organisation functions – Planning and construction[8].

Main Functions	Partial Functions	Sub Functions
Planning and construction functions	Planning and design	❑ Resource management and water protection ❑ Water supply master plan with medium and long term demand and supply and technical objectives ❑ Asset management planning and rehabilitation programmes ❑ Planning, design and calculation of plants, networks and equipment ❑ Specification of plants, construction and materials selection ❑ Planning and design of technologies e.g., information technologies ❑ Procurement of external design services in relation to the above ❑ Statistics
	Construction	❑ Procurement of construction works by external contract ❑ Direction of works undertaken by direct labour ❑ Supervision, quality control and acceptance ❑ Commissioning of new plants, networks and equipment ❑ Payments and financial control

[8] Planning and construction is the set of tasks related to planning and provision of new assets and reinforcement, replacement and renovation of existing assets.

Table 67. Organisation functions – Operations and maintenance.

Main Functions	Partial Functions	Sub Functions
Operations and maintenance functions	Operations	❑ Supervision of source protection areas ❑ General system monitoring and control ❑ Standby duties and risk management ❑ Documentation, data collection, record keeping and data processing (including management and operation of information systems) ❑ Water quality monitoring, sampling and analysing ❑ Water loss management ❑ Operational acceptance of new plants, networks and equipment ❑ Re-commissioning of systems after shut down
	Maintenance and repair	❑ System inspection ❑ Regular maintenance checks or service activities ❑ *Repairs* of failures and other defects ❑ Cleaning ❑ *Refurbishment* ❑ Regular meter replacement and maintenance ❑ Maintenance and *repair* of fire fighting assets (whenever this is a responsibility of the *utility*)

The technical main functions – planning, construction, operations and maintenance – can also be split in a different way, from the perspective of the typical human resources allocation:

- Water resources and catchment management;
- Abstraction and treatment;

- Transmission, storage and distribution;
- Water quality monitoring;
- Meter management;
- Support services (central stock repository, central workshops and central vehicle fleet).

1.3 Financial definitions (*see also part i, point 5.4, economic and financial indicators*)

This section presents the financial definitions for the annual profit and loss architecture, the investment definitions, the annual changes in financial position and the structure of the balance sheet at the end of the year.

The annual profit and loss definitions presented in Table 68 and Table 69 refer to financial accounting principles. In the case of multi-service utilities, the profits and losses related to the water service (including the technical and the non-technical activities) are specifically identified by management accounting.

Previously, widely accepted financial accounting principles were used, notwithstanding some variations between countries. More recently comparing methodologies at an international level has grown, as a consequence of growing international shareholding and trade. There is a real need for International Financial Reporting Standards (IFRS), designed as a common global language for business affairs to make company accounts understandable and comparable across international boundaries. Financial statements need to reflect a true and fair view of the financial position of an organization and are used by various professionals and regulators. These statements enable them to check the financial position of a business for a specific period.

Currently most countries of the world require or permit IFRS reporting. IASB have been progressively replacing many different national accounting standards and have established the rules that accountants must follow in order to maintain accounts that are comparable, understandable, reliable and relevant for internal or external users.

All over the world, utilities exist that operate public services under concession contracts. To properly harmonize various accounting aspects of this particular situation the International Financial Reporting Interpretation Committee issued, in November 2006 and under the IASB umbrella, the official interpretation "IFRIC 12 – Service Concession Arrangements" which became effective on January 1st 2008.

Recent past experiences prove that the IWA PI Systems have been adopted worldwide. Given the current steady trent, adoption is expected to increase in the near future. Given the adoption of IAS compliant Accounting Systems highly influences the main financial variables and PIs, comparing PIs of different utilities may become less meaningful if not all of them have gone in this direction. Comparison based on the same concepts and definition is needed in order to be valid.

So this section of the Manual will focus how IAS adoption can modify (both in concept and in value) some variables allocated into the architecture of both the Profit and Loss and the Balance Sheet at the end of the fiscal year.

Arrangements will thus be identified to properly "adjust" the economic and financial variables to be considered in order to preserve the validity and benefits of the IWA PI's System and consistent detailed definitions will be provided.

Some of these variables are relevant by themselves to support management, although they are not performance indicators as defined in this IWA PI MBP.

Detailed definitions related to Table 68 are given below:

Table 68. Annual profit and loss architecture.

Annual costs (–G4 = G5 + G6)

Running costs (G5 = Operating costs – G35)

Capital costs (G6 = G28 + G31)

Sales revenues (G3) Work in progress (-) Capitalised costs of self-constructed assets (G35) Other operating revenues (-)	Operating revenues (G2)
Operational costs* (G7): ⇒ External services* (G9) ⇒ Imported (raw and treated) water* (G10) ⇒ Energy* (G11) ⇒ Chemicals, other consumables and materials* (G12) ⇒ Leasing and rentals* (G13) ⇒ Taxes, levies and fees* (G14) ⇒ Exceptional earnings and losses* (G15) ⇒ Other operating costs* (G16) Internal manpower costs* (G8)	Operating costs
E.B.I.T.D.A. (G54 + G28) Earnings before interests, taxes, depreciation and amortisation	
Depreciation (referred to the book values) (G28)	
E.B.I.T. = O.I. (G54 = G2–G7–G8–G28) Earnings before interests and taxes = Operating Income	
Interest expenses (G29) Interest income (G30)	Net interest (G31 = G30–G29)
E.B.T. = G.I. Earnings before taxes = Gross Income (-)	
Taxes (-) NET INCOME (G56)	

* Gross value; the variable between brackets is the corresponding net value, i.e., not including the contribution for capitalised cost of self-constructed assets.

Table 69. Detail definitions related to Table 68.

Item in Table 68	Include
IMPORTED (RAW AND TREATED) WATER	BULK SUPPLY IMPORTS: total payments, for imported bulk supplies (imported raw water and/or imported treated water).
ENERGY	All electricity costs related to the water supply service, including energy for pumping and for other water supply *utility* activities (e.g., workshops, office buildings energy consumption, laboratories).
EXTERNAL SERVICES	OUTSOURCING: outsourcing of technical or administrative services, such as consultants, contractors, *utility* operational tasks, meter reading and accounting fees. SOFTWARE LICENCES AND IT SUPPORT: licence fees on computer software and technical support by software companies. ASSOCIATED COMPANIES: costs of associated companies that are not included in other items. THIRD PARTY SERVICES: operating costs of providing water services to third parties (other than the regulated water supply function) that are not included in other items.
LEASING AND RENTALS	Payments for leasing or renting premises, vehicles, mobile and fixed plant and equipment.
PURCHASES	MATERIALS, CHEMICALS AND OTHER CONSUMABLES: all materials, chemicals and other consumables apart from electrical energy, that are not in HIRED AND CONTRACTED SERVICES and which are required for the operation of sources, treatment plants, transmission and distribution systems or for the non-technical functions (e.g., office consumables).
TAXES, LEVIES AND FEES	Any operating licence paid to a governmental or municipal authority other than income taxes (e.g., abstraction charges, local authority rates).
EXCEPTIONAL EARNINGS AND LOSSES	Any exceptional income or expenditure from donations, investment subsidies, compensations or adjustments related to sales/writing-off of fixed assets.

Table 69. Detail definitions related to Table 68 (Contd.).

Item in Table 68	Include
OTHER OPERATING COSTS	OTHER DIRECT COSTS: any other operating costs (but excluding interest and taxation, on an aggregated basis). GENERAL AND SUPPORT EXPENDITURES: the aggregate direct cost of GENERAL AND SUPPORT ACTIVITIES (manpower costs excluded) (see section 1.2 for definitions). CUSTOMER SERVICES: costs directly associated with customer services that are not included in previous items, related to customer accounting, reading of meters, debt recovery, costs of disconnections, customers' enquiries and complaints handling. SCIENTIFIC SERVICES: costs directly associated with scientific and laboratory services and with the monitoring of quality that are not included in previous items. OTHER BUSINESS ACTIVITIES: costs directly associated with other business activities that are not included in previous items, except for cost depreciation. DOUBTFUL DEBTS: charge/credit to the profit and loss account for bad and doubtful debts.
INTERNAL MANPOWER COSTS	EMPLOYMENT COSTS: the sum of the total manpower costs of permanent and temporary personnel directly employed by the *utility*, including employment- related social costs and benefits paid by the employer.
CAPITALISED COSTS OF SELF-CONSTRUCTED ASSETS (negative allocation)	The summation of the amounts in each of the above cost categories that have been incurred in the construction of new or rehabilitated assets (including, in the case of concessions, the assets constructed, upgraded or rehabilitated according to the concession arrangement). *For running cost PI calculation purposes, capitalised costs of self-constructed assets are to be apportioned negatively against each of the above running cost items having due regard to the strict observance of the following relationship:* Running costs from the Annual Profit and Loss Architecture = Operation and Maintenance Costs + Internal Manpower Costs – Capitalised costs of self-constructed assets =

Table 69. **Detail definitions related to Table 68 (Contd.).**

Item in Table 68	Include
	Net value of: (Imported waters + Energy + External services + Purchased merchandises + lease rentals + operating expenditures + internal manpower costs)
DEPRECIATION (ON BOOK VALUES)	COST DEPRECIATION: cost depreciation charge on water service fixed assets AMORTISATION OF INTANGIBLE ASSETS: any amortisation or other reduction in the balance sheet valuation of intangible assets. THIRD PARTY SERVICES: cost depreciation on assets relating to third party services, together with any infrastructure renewal charge for infrastructure assets relating to third party services.
NET INTEREST	NET INTEREST: Net cost of short, medium and long-term loan capital (INTEREST EXPENSES – INTEREST INCOME).
INCOME TAXES	TAXES: All taxes and levies on gross income related to water supply activities. *Tax costs and levies strictly connected with plants' operations (such as sewerage charges on treatment wastes, charges for water abstraction, pipeline and concession charges, environmental levies, water control authority charge etc.) have to be regarded as operational costs and included in* TAXES, LEVIES AND FEES (RUNNING COSTS).

Table 70. **Investment definitions.**

Term	Definition
INVESTMENTS SUBJECT TO DEPRECIATION	Investments that have to be subject to depreciation as listed in detail by the general accepted accounting principles.
DEPRECIATED HISTORICAL VALUE OF WATER SERVICE ASSETS	Sum, over all the water service assets, of the depreciation value applied since the date of purchase.
HISTORICAL VALUE OF WATER SERVICE ASSETS	Refers to the gross book value of total expenditures (that appear for the net value in Table 72) in infrastructure and non-infrastructure water service assets.

Table 71. Annual changes in financial position.

Application of Funds	Sources of Funds
Capital expenditures (Investments)[9]: ⇒ Intangible assets ⇒ licences and rights	Internal sources from ***CASH FLOW:*** ⇒ retained earnings (after dividends and other payments out of net income) ⇒ depreciation
⇒ Water service assets ⇒ pipeline network ⇒ plants and machinery ⇒ other equipment	External sources: ⇒ capital receipts ⇒ borrowing and loans obtained ⇒ sale of fixed assets ⇒subsidies to the investment
⇒ Financial assets ⇒ shares in other companies ⇒ other financial investments **Financing:** ⇒ Borrowings and loans repaid	

Working Capital increase[10]	Working Capital decrease[10]

The above tables refer to the financial accounting principles as defined in the general doctrine and generally accepted, even if with some differences between countries (GAAP).

The transition towards IFRS creates significant changes to the financial reporting of companies, including changes in measurement, disclosure and presentation.

The adoption of standards showed that changes in the measurement of financial items significantly affected valuations and values, with particular attention to fair value disclosure, use of provisions and asset value impairments.

[9] Annual capital expenditure (investment in tangible, intangible and financial assets) is the same as "total annual investment".

[10] The increase or the decrease is the balance between application and sources.

Table 72. Structure of the balance sheet at the end of the year.

	Assets	**Shareholders' Equity and Liabilitie**	
Total assets (G50)	Intangible assets ⇒ *goodwill* ⇒ net value of licences and rights	Shareholders' Equity (G48) ⇒ subscribed share capital ⇒ capital reserves ⇒ other reserves ⇒ net income for the year	
	Water service assets (G42-G41) ⇒ net value of *water utility* plants ⇒ net value of other assets	Long term liabilities (G52) ⇒ bonds ⇒ Differed revenues (subsidies to the investment) ⇒ long term financial debts	Total debt G47 = G52 + G53
	Financial assets ⇒ net value of financial investments		
	Current assets[11] (G49) ⇒ cash at bank and in hand ⇒ accounts receivable from drinking water (G38) ⇒ other accounts receivable ⇒ inventories (G51) ⇒ prepaid expenses	Current liabilities[11] (G53) ⇒ accounts payable ⇒ current portion of long term debt ⇒ miscellaneous current liabilities	

The adoption of IFRS principles highly influences main financial variables and reports and analysts need to distinguish reported performance changes caused by the transition to IFRS from those caused by changes in the business.

It is important to remark that changes coming from IFRS adoption do not influence the components of the involved ratios or the rationales the same ratios are based on.

Changes from IFRS

The current literature comprehensively covers IFRS and this can be referred to for deeper analysis. The major interest, in this Manual, is to observe in detail

[11] Current assets – Current liabilities = Working capital

the impact of IFRS on financial indicators and variables that are considered in the IWA PI's System.

First of all, it is useful to identify the most relevant impacts and changes on the Profit and Loss and Balance Sheet architectures on which the Performance Indicators are based.

Given that a "schematic" approach seems to be a better fit for such a purpose, the Financial Report IFRS compliance compared with National GAAP traditional reports shows potential changes apply differently to water utilities and this can be observed in Table 73 and Table 74.

The Annual Profit and Loss and Balance Sheet architectures, as designed in previous Table 68 and Table 72, are the same schemes. Values are allocated in terms of normal proportions in the water sector.

The big picture

From a theoretical perspective a change in accounting should not change the value of a company, apart from potential changes in behaviour that the changes may provoke. For instance, fair value accounting could lead to potentially suboptimal risk-approaching strategies in order to reduce P&L volatility.

However, in practice the adoption of IFRS, and in particular the potential new disclosures and the change in measurements, may ultimately affect the way interested stakeholders value companies:

- *Measurement*: accounting items may often be subject to new measures; for instance revenues are to be recognised in a different way or the value of certain assets or liabilities has to be marked to market.
- *Disclosure*: some required additional information is to be disclosed, or to be disclosed in a different way.
- *Presentation*: information is currently presented differently; for instance, extraordinary items may no longer exist, even though the companies may have chosen to show the impact from exceptional items separately.

The above changes can be summarized, for a better understanding, by the following Table 75.

Table 73. Comparative Profit and Loss.

K€	IFRS	GAAP **	Changes
Sales revenues (G3)	10,000	10,000	Changes in revenue recognition, e.g., mark to market of not 'own use' may generally not apply to Water Utilities
Work in progress (-)			
Capitalised costs of self-constructed assets (G35)	2,150	6,150	Changes in cost capitalisation, acquisition, maintenance & construction of infrastructures under concession
Other operating revenues (-)	1,990	1,990	Changes in revenue recognition, may generally not apply to Water Utilities
Operational costs (G7):			
External services (G9)	2,500	6,500	Changes in cost capitalisation, acquisition, maintenance & construction of infrastructures under concession
Imported (raw and treated) water (G10)	150	150	
Energy (G11)	1,350	1,350	
Chemicals, other consumables and materials (G12)	3,390	3,390	
Leasing and rentals (G13)	650	650	

Table 73. Comparative Profit and Loss (Contd.).

K€	IFRS	GAAP **	Changes
Taxes, levies and fees (G14)			
Exceptional earnings and losses (G15)	(30)	(30)	
Other operating costs (G16)	1,140	1,120	Changes in revenue recognition that may generally not apply to Water Utilities, stricter definition of provisions
Internal manpower costs (G8)	2,540	2,580	Changes in cost capitalisation, restructuring charges (Fair value of termination indemnity and other benefits to personnel)
E.B.I.T.D.A. (G54 + G28) Earnings before interests, taxes, depreciation and amortisation	2,450	2,430	Increased volatility due to fair value accounting. Some charges from depreciation to expense
Depreciation (referred to the book value) (G38)	1,970	1,980	Changes in useful lives and estimate of value, goodwill amortisation disappeared since 2004, yearly impairments instead
Operating income (G54 = G2-G7-G8-G28) *Earnings before interests and taxes = Operating Income*			Increased volatility due to fair value accounting and depreciation

DEFINITIONS

Table 73. Comparative Profit and Loss (Contd.).

K€	IFRS	GAAP **	Changes
Interest expenses (G29)	320	290	Impact from fair value accounting and amortisation of debt raising costs. debt definition. Fair value of derivatives.
Interest income (G30)	10	30	Impact from fair value accounting, credit definition.
E.B.T. = G.I. *Earnings before taxes = Gross Income*	170	190	Increased volatility
Taxes (-)	150	140	Changes in tax estimations, including deferred tax. Potential tax impacts of fair value accounting
NET INCOME (G56)	20	50	Significant volatility

* Values don't reflect any real company situation

Impact on Financial Ratios

Previous Table 73 *and* Table 74 show potential changes the IFRS adoption is bringing to water utilities financial statements, according to the key issues of related IFRS principles described in Table 75, so highlighting the problem to establish the level of peers comparability when some apply IFRS and others continue to use national standards.

Case studies and literature indicate that the image of listed companies differs significantly when IFRS rather GAAP are applied in the preparation of financial information. This is the reason why understanding the rationales of such differences is important to assess the impact of IFRS on calculation and volatility of financial ratios.

Table 74. Comparative balance sheet at the end of the year.

K€	IFRS	GAAP **	Changes
ASSETS			
Intangible assets *goodwill* *net value of licences and rights*	 2,290 19,570	 1,730 19,570	Yearly impairment tests instead of yearly depreciation Changes in cost capitalisation, acquisition, maintenance & construction of infrastructures under concession which may reflect more intangible assets
Water service assets (G42-G41) *net value of water utility plants* *net value of other assets*	1,950	1,950	Above changes in cost capitalisation may reflect less water service assets
Financial assets *net value of financial investments*	380	360	Changes in consolidation. Fair value accounting.
Current assets[1] (G49) *cash at bank and in hand* *accounts receivable from drinking water G38)* *other accounts receivable* *inventories (G51)* *prepaid expenses*	 550 21,060 2,680 170	 550 21,170 2,680 170	Fair value accounting
Total Assets (G50)	48,650	48,180	
SHAREHOLDERS' EQUITY AND LIABILITIES			
Shareholders' Equity (G48)	15,500	15,500	

Table 74. Comparative balance sheet at the end of the year (Contd.).

K€	IFRS	GAAP **	Changes
subscribed share capital *capital reserves* *other reserves* *net income for the year*	 3,190 20	 2,220 50	Increased volatility from fair value accounting and adjustments to results.
Long term liabilities (G52) *bonds* *deferred revenues (subsidies to the investment)* *long term financial debts*	 3,450 1,150	 3,470 1,600	Fair value accounting. Discounting of provisions for risks & expenses. Potential change in value & classification.
Current liabilities (G53) *accounts payable* *current portion of long term debt* *miscellaneous current liabilities*	 14,830 2,700 7,810	 14,830 2,700 7,810	
Total Liabilities (G50)	48,650	48,180	

* Values don't reflect any real company situation

**GAAP reflects IWA PI's System interpretation of commonly accepted national accounting principles.

It has been seen that changes are most significant for those figures that are more relevant for the evaluation of a financial structure such as debtors, cash and cash equivalents, equity, long-term and total liabilities. So when analysing top significant adjustments to adapt financial statements to international standards, the main reasons of variation, can be summarised as follows:

- the application of fair value to financial instruments, the reclassification of accounts, and changes in the scope of consolidation impact on current assets;
- change in the rules for the valuation of debts and changes in the scope of consolidation directly influence liabilities, while direct adjustments and the indirect effect of adjustments to results impact on equity figures;

Table 75. IAS/IFRS key issues.

IAS/IFRS	Key issues
Fair value accounting (standards IAS 39 and IAS 32):	IAS 39 (Financial Instruments: Recognition and Measurement) outlines the requirements for the recognition and measurement of financial assets and liabilities and some contracts to buy or sell non-financial items. IAS 32 (Financial Instruments: Presentation) outlines the accounting requirements for the presentation of financial instruments, particularly as to their classification into financial assets, financial liabilities and equity instruments.. The above standards apply to all types of financial instruments, except for those which are specifically scoped out. Financial instruments are initially recognised when an entity becomes a party to the contractual provisions of the instrument, and are classified into various categories depending upon the type of instrument, which then determines the subsequent measuring methodology. Fair value is the amount for which an asset could be exchanged, or a liability settled, between knowledgeable, willing parties in an arm's length transaction. Connected value changes are recognised in profit or loss. Common example of financial instruments within the scope of IAS 39 are: cash, demand and time deposit, commercial paper, receivables and payables, debt and equity securities, derivatives and so on. Fair value accounting likely brought significant volatility to the P&L (and to the balance sheet). Nonetheless literature shows that this volatility seems more likely to affect utilities with substantial exposure to commodities and thus to occur in non-regulated businesses such as generation and supply of electricity, or gas imports and supply, rather than in fully regulated businesses such as gas transmission and water supply.

- *global fixed assets and inventories seem not to vary significantly*: the former thanks to the generally adopted option not to change the fixed asset valuation criterion applied (acquisition cost) while the latter depending

Table 75a. IAS/IFRS key issues.

IAS/IFRS	Key issues
Pensions and benefits for employees (standard IAS 19):	IAS 19 (Employee Benefits) outlines the accounting requirements for employee benefits, including short-term benefits (e.g., wages and salaries, annual leave), post-employment benefits such as retirement benefits, other long-term benefits (e.g., long service leave) and termination benefits. The standard establishes the principle that the cost of providing employee benefits should be recognised in the period in which the benefit is earned by the employee, rather than when it is paid or payable, and outlines how each category of employee benefits is measured, providing detailed guidance in particular about post-employment benefits. The standard separately rules short term benefits and long term benefits or contribution plans, but generally speaking when the payment obligations do not fall due within 12 months after the end of the period in which the employee renders the service, it should be discounted to their present value. The amount thus recognised in the balance sheet should represent the present value of expected future payments required to settle the obligation resulting from employee service in the current and prior periods as adjusted for unrecognised actuarial gains and losses and unrecognised past service cost, and reduced by the fair value of plan assets at the balance sheet date. The fair value of pension assets and actuarial liabilities represented into the balance sheet may increase the volatility for the income statement depending on both the adopted actuarial approach and the criteria the local GAAP previously followed to allocate these obligations.

on the evaluation method previously adopted. Nevertheless significant variations can be observed when separately analysing water service and intangible fixed assets due to changes in cost capitalisation and in goodwill valuation;

Table 75b. IAS/IFRS key issues.

IAS/IFRS	Key issues
Interest expenses (Standard IAS 23)	IAS 23 (Borrowing Costs) requires that borrowing costs directly attributable to the acquisition, construction or production of a 'qualifying asset' (one that necessarily takes a substantial period of time to get ready for its intended use or sale) are included (capitalised) in the cost of the asset. Other borrowing costs are recognised as an expense. Eligible assets could be property, plant, and equipment and investment property during the construction period, intangible assets during the development period, or even inventories. Borrowing costs include interest on bank overdrafts and borrowings, amortisation of discounts or premiums on borrowings, finance charges on finance leases and exchange differences on foreign currency borrowings where they are regarded as an adjustment to interest costs. IFRS 23 thus imply some change (increase) to the carrying amount of assets and correspondent higher depreciation that reflect increasing Profit and Loss volatility.

Table 75c. IAS/IFRS key issues.

IAS/IFRS	Key issues
Declining value of Assets (standards IAS 36): Goodwill	IAS 36 (Impairment of Assets) seeks to ensure that an entity's assets are not carried at more than their recoverable amount (i.e., the higher of fair value less costs of disposal and value in use). Except for some assets that other IAS specifically deal with, this standard applies to all assets including goodwill and intangible assets acquired in business combinations.

Table 75c. IAS/IFRS key issues (Contd.).

IAS/IFRS	Key issues
	At the end of each reporting period, an entity is required to assess whether there is any indication that an asset may be impaired (i.e., its carrying amount recognised in the balance sheet may be higher than its recoverable amount, which is the higher between its fair value less costs of disposal and its value in use, which is the present value of the future cash flows expected to be derived from). IAS 36 has a list of external and internal indicators of impairment. If there is an indication that an asset may be impaired, then the asset's recoverable amount must be calculated. Nevertheless the recoverable amounts of goodwill and certain detailed intangible assets are measured annually, whether or not there is any indication that it may be impaired. The above thus means that under IFRS, goodwill ceases to be systematically amortised in the Profit &Loss: past acquisitions will most likely not be reaccounted for, and hence the goodwill will stay in the balance sheet and goodwill will have a yearly impairment test. To test for impairment, goodwill must be allocated to each of the acquirers' cash-generating units, or groups of cash-generating units, that are expected to benefit from the synergies of the combination, irrespective of whether other assets or liabilities of the acquiree are assigned to those units or groups of units. It can be desumed that the impairment tests IAS 36 requirements are more likely to bring water utilities and market attention to the value of assets and, more in particular, goodwill can be desumed as the asset that is likely to be most tested for impairment. Given that IFRS is, in general, stricter about the expenses and the maintenance costs that can be capitalised the accounting experts feel that this seems likely to result in lower capital expenditure, higher expenses and lower depreciation, also connected with changed concept of useful lives.

Table 75d. IAS/IFRS key issues.

IAS/IFRS	Key issues
Intangible Assets accounting (standard IAS 38 and IFRIC 12)	IAS 38 (Intangible Assets) outlines the accounting requirements for intangible assets, which are identifiable non-monetary assets without physical substance (either being separable or arising from contractual or other legal rights) that are controlled by the entity as a result of past events (for example, purchase or self-creation) and from which future economic benefits are expected. The pattern of expected benefit directly reflects the asset useful life so remarkably changing the GAAP concept of useful life and consistently impacting on assets depreciation. Intangible assets meeting the relevant recognition criteria are initially measured at cost and subsequently measured at cost (less accumulated amortization and impairment losses) or using the revaluation model (based on fair value), and amortised on a systematic basis over their finite useful lives. In case of indefinite useful life the intangible should not be amortised. In any case the asset should also be assessed for impairment in accordance with IAS 36.
Intangible Assets accounting (The IFRIC 12	All over the world, utilities exist that operate public services under concession contracts. To properly harmonize various accounting aspects of this particular situation the International Financial Reporting Interpretation Committee issued, in November 2006 and under the IASB umbrella, the official interpretation "IFRIC 12 – Service Concession Arrangements" which became effective on January 1st 2008. IFRIC 12 defines a service concession arrangement as an arrangement whereby a government or other public sector body contracts with a private operator to develop (or upgrade), operate and maintain the grantor's infrastructure assets such as roads, bridges, tunnels, airports, energy, water and gas distribution networks, prisons or hospitals. The grantor controls or regulates what services the operator must provide using the assets, to whom, and at what price, and also controls any significant residual interest in the assets at the end of the term of the arrangement.

Table 75d. IAS/IFRS key issues (Contd.).

IAS/IFRS	Key issues
	IFRIC 12 draws also a distinction between two types of service concession arrangement. In one, the operator receives a financial asset, specifically an unconditional contractual right to receive a specified or determinable amount of cash or another financial asset from the government in return for constructing or upgrading a public sector asset, and then operating and maintaining the asset for a specified period of time. This case of concession arrangements, known as BOT – Built, Operate & Transfer – contracts, is in the field of large public infrastructures like highways or dams. In the other case, the operator receives an intangible asset, specifically a right to charge for use of a public sector asset that it constructs or upgrades and then must operate and maintain for a specified period of time. This case of concession arrangements is in the field of local public services like water and gas. Operating revenues (coming from both financial and intangible received asset) and operating costs (required to manage the received asset) are measured at fair value for each concession arrangement item and basing on the concession arrangement progress (component approach). Constructed assets, both financial and intangible, are measured at their fair value in the Profit and Loss statement and according the type of the concession arrangement: financial assets will follow the IAS 32 (Financial instruments) whilst intangible assets will follow the IAS 38 (Intangible assets). In practice the above means that in the IFRS context the assets constructed or upgraded according to the concession arrangement are represented in the Profit and Loss Statement and get the same approach of the self-constructed assets: paid (operating) costs are annually capitalized and allocated among interested assets (Intangible or Financial).

Table 75d. IAS/IFRS key issues (Contd.).

IAS/IFRS	Key issues
	Remembering that in the GAAP context the assets constructed or upgraded according to the concession arrangement were generally represented as "investments" (or capex) directly allocated among interested assets and without any involvement of Profit and Loss statement, IFRS adoption reflect, thanks to the IFRIC 12 interpretation, increased operating costs and more correspondent capitalisation which do not have any influence on EBITDA volatility. Received and increased intangible assets will be then amortised on a systematic basis over the duration of the concession

- in the income statement, significant variations can be observed in operating income, that reflect increasing volatility of the bottom line, depending on potential differences in the treatment of revenues and expenses (cost capitalisation, asset impairment, fair value accounting etc.).

Such differences in the measurement of accounting figures under IFRS rather than traditional GAAP may directly affect ratios calculation influencing the numerator, the denominator, or both.

Where the difference in measurement affects only the numerator or only the denominator, the effect of changes is straightforward, easy to identify and to interpret. For example, the current ratio is higher under IFRS (everything else being equal) if current assets are higher but current liabilities remain unchanged.

Identification and interpretation is less obvious in cases of numerous diverging effects on ratios. For example, a lower profit under IFRS will pull down the return on assets (ROA) by reducing the numerator but, at the same time, will pull it up by reducing the denominator.

Moreover, there might be distinct accounting differences between IFRS and traditional GAAP that have opposite effects on a particular ratio. An

Table 75e. IAS/IFRS key issues.

IAS/IFRS	Key issues
Deferred tax (standard IAS 12)	IAS 12 (Income Taxes) implements a so-called 'comprehensive balance sheet method' of accounting for income taxes which recognises both the current and the future tax consequences of the future recovery or settlement of the carrying amount of an entity's assets and liabilities. Differences between the carrying amount and tax base of assets and liabilities, and carried forward tax losses and credits, are recognised, with limited exceptions, as deferred tax liabilities or deferred tax assets, with the latter also being subject to a 'probable profits' test. The general principle is, on the one hand, that deferred tax liabilities should be recognised for all taxable temporary differences (which are intended as the difference between the carrying amount of an asset or liability and its tax base) whilst, on the other side, a deferred tax asset should be recognised for deductible temporary differences, unused tax losses and unused tax credits to the extent that it is probable that taxable profit will be available against which the deductible temporary differences can be utilised, unless the deferred tax asset arises from. The temporary difference approach used in the IFRS context is, beyond any doubt, wider than the timing approach generally adopted in the GAAP context and may result in increases both in deferred tax assets and liabilities, particularly when fair value accounting and impairing methodologies significantly impact on Profit and Loss.

example is the impact on the current ratio of higher current assets under IFRS due to fair value accounting of revenues receivables concurrent with higher liabilities due, for example, to the recognition of a finance lease liability.

However and generally speaking it can be gathered that solvency and indebtedness ratios, as well as the return on assets and return on equity can vary significantly as a result of the changes in the balance sheet and income statement.

Operating financial variables continue to be useful proxies for cash but the adoption of actualization methodologies are bringing new concepts of margins; key figures such as debt, pensions, provisions or minorities, change substantially and the changes in the figures themselves (be it Earnings Before Interest, Taxes, Depreciation, and Amortization, (EBITDA) or Net debt as a result of fair value accounting and cost capitalization policies), jointly with the likely increase in Profit & Loss volatility, may affect the overall perception of relative and absolute company value.

The increased volatility of net income makes EPS (Earning per Share) related ratios of even less use than previously was for largest utilities comparable ratios (including Price Earning Ratio, (PER) or Earning per Share, (EPS) growth).

It can be concluded that the economic and financial positions of water utilities reflected in accordance with IFRS appear significantly different from the image presented by local accounting standards, and into a general comparison framework characterized by an increased disclosure providing new and more information. This may enable the identification of the potential sources of underperformance earlier (for instance, in case of potentially stricter impairment rules).
(Nota HA: parece-me dispensável

The number of financial ratios impacted by IFRS adoption is considerable beyond any doubt, but it's anyway important to remark once more that such an impact doesn't influence neither the components of the involved ratios nor the rationales the same ratios are basing on.

What is changing is the output of the ratios, that is the quality of the provided information and thus the ending result of the peers comparison, particularly when some among those peers in the panel don't adopt IFRS principles.

Should this last an unavoidable option it is similarly unavoidable to prior make uniform the data base of the panellists by reconverting the values represented into the Profit and Loss and Balance Sheet architectures from IFRS in GAAP

Table 76. Impact on financial indicators (P&L).

Cost group	Variables			Changes	Involved Fi's: Which	Involved Fi's: Where*
	Sales revenues (G3) Work in progress (-) Other operating revenues (-)	Total Revenues (G1)	Operating revenues (G2)	Y	Fi38	Den
Running costs (G5 = Operating costs – G35)	Capitalised costs of self-constructed assets (G35)					
Annual costs (G4 = G5 + G6); Running costs (G5 = Operating costs – G35)	Operational costs (G7): ⇒ External services (G9) ⇒ Imported (raw and treated) water (G10) ⇒ Energy (G11) ⇒ Chemicals, other consumables and materials (G12) ⇒ Leasing and rentals (G13) ⇒ Taxes, levies and fees (G14) ⇒ Exceptional earnings and losses (G15) ⇒ Other operating costs (G16)		Operating costs	Y Y Y	Fi4, Fi5, Fi7, Fi8, Fi9, Fi10, Fi11, Fi12, Fi13, Fi14, Fi15, Fi16, Fi17, Fi18, Fi19, Fi20, Fi21, Fi22, Fi30, Fi31, Fi47	Num Both Den
	Internal manpower costs (G8)			Y		
Running costs (G5 = Operating costs – G35)	E.B.I.T.D.A. (G54 + G28) Earnings before interests, taxes, depreciation and amortisation			Y		
Capital costs (G6 = G28 + G31)	Depreciation (referred to the book values) (G28)			Y	Fi4, Fi6, Fi36, Fi42, Fi44	Num Both Den
Annual costs (G4 = G5 + G6)	E.B.I.T. = O.I. (G54 = G2 – G7 – G8 – G28) Earnings before interests and taxes = Operating Income			Y	Fi23, Fi24,	
Capital costs (G6 = G28 + G31)	Interest expenses (G29) Interest income (G30)		Net interest (G31= G30 – G29)	Y	Fi30, Fi33	
	E.B.T. = G.I. (G54) Earnings before taxes = Gross Income (-)			Y	Fi42, Fi44	Num
	Taxes (-)			Y		
	NET INCOME (G56)			Y	Fi43	Num

* Where changes impact in Fi calculation (Den: Denominator, Num: Numerator, Both)

Table 77. Impact on financial indicators (BS).

Variables	Changes	Involved Fi's	
		Which	Where*
ASSETS			
Intangible assets *goodwill* *net value of licences and rights*	Y	Fi25 (G32), Fi26 (G32, G33), Fi27 (G32, G34), Fi34 (G32)	Both
Water service assets (G42-G41) *net value of water utility plants* *net value of other assets*			
Financial assets *net value of financial investments*	Y		
Current assets[1] (G49) *cash at bank and in hand* *accounts receivable from drinking water G38)* *other accounts receivable* *inventories (G51)* *prepaid expenses*	Y	Fi32 (G38), Fi38 (G51) Fi41 (G49, G53)\	Num Both
Total Assets (G50)	Y	Fi44,Fi45	Den
SHAREHOLDERS' EQUITY AND LIABILITIES			
Shareholders' Equity (G48) *subscribed share capital* *capital reserves* *other reserves* *net income for the year*	 Y Y	Fi40 (G47, G48), Fi43 (G48)	Both Den
Long term liabilities (G52) *bonds* *deferred revenues (subsidies to the investment)* *long term financial debts*	 Y Y		
Total Debts (G47)			
Current liabilities9 (G53) *accounts payable* *current portion of long term debt* *miscellaneous current liabilities*			
Total Liabilities (G50)	Y		

* Where changes impact in Fi calculation (Denominator, Nominator, Both)

Table 78. From IFRS to GAAP: adjustments.

Criteria	Involved Variables in the P&L and BS	Adjustments IFRS vs/ GAAP
Fair value application	Interest income Financial assets Account receivable Long term liabilities	Discount vs/ Nominal value Discount vs/ Nominal value Discount vs/ Nominal value Senior debt interest capitalization & Provisions to personnel discount vs/ debt increase
Benefits to personnel	Internal manpower costs	Benefits & Termination indemnity for the year discount vs/ nominal cost
Cost capitalization	Interest expenses Goodwill	Capitalization vs/ P&L allocation Impairment vs/ Depreciation
Depreciation	Assets Useful Life Goodwill	Water service vs/ Intangible Impairment vs/ Depreciation
Equity	Reserves Net income for the year	direct adjustments and the indirect effect of adjustments to results
Taxes	Deferred taxes	temporary difference approach vs/ timing approach

criteria, like for example do in practice the Holdings not yet IFRS compliant needing to consolidate the results of subsidiaries IFRS compliant into the Group Financial Report.

This backwards procedure will basically involve the application of fair value to financial instruments and benefits to personnel, the rules adopted for the valuation of debts, the differences incurred in cost capitalisation, depreciation and asset impairment (with particular reference to goodwill treatment) as detailed in following Tab. Jjj.

As for the Fixed Assets it is important to remember that in the IFRS context (as better defined by the above IFRIC 12 interpretation, see previous Tab. Zzz), Water Utilities represent the assets constructed or upgraded according the concession arrangement in the Profit and Loss Statement, like the self-constructed assets, and annually capitalize paid (operating) costs through allocation among intangible assets to be depreciated, on a systematic basis, over their finite useful lives, that is the duration of the concession.

Differently, in the GAAP context the assets constructed or upgraded according the concession arrangement are generally represented as "investments" (or capex) directly allocated among interested assets (normally tangible) without any involvement of Profit and Loss statement.

It comes from the above that IFRS adoption reflect increased operating costs and more correspondent capitalisation but this is neutral on the bottom line.

On the other side the IWA PI's System explicitly excludes self-constructed assets from any indicator (see this Manual Part II, chapter 2 and 3, Performance Indicators and Variables Definitions), so such an increase of both operating costs and correspondent revenues is relevant just for the investments analysis (Fi25, Fi26, Fi27, Fi34, Fi35, Fi36, Fi42) that in the IFRS context reveals a predominance of Intangible Assets.

In some GAAP context a similar repositioning among the intangible assets of the assets underlying the concession contract is however to be observed so the above Performance Indicators and connected variables have been properly adjusted in following sections 2.8 and 3.7 in this Part II.

Less evident, but directly connected, impact is finally to be observed on Depreciation and in connection with different concepts (IFRS vs/ GAAP) of the useful lives between tangible and intangible assets.

In conclusion it can be assessed that the IFRS adoption in Water Utilities have considerable impact on the values as represented into the Financial Report, so inducing lot of changes in both the quality and the rationales of the provided financial information.

In the framework of performance comparisons this impact doesn't nevertheless operate in case of internal comparison during time, given that First Time Adoption procedures (as defined into the interpretation SIC8) rule the application of transitional provisions in the period of first-time application of IASs so that the pro-forma financial statements should be prepared and presented as it had always been prepared in accordance with the IASs.

On the contrary a prior investigation of adopted accounting principles and, if the case, a prior harmonization of panellists data base are thus to be strongly recommended as pivotal peculiarities of any comparing panel careful selection when comparisons is developed with both national and international peers Utilities

1.4 Complementary definitions

In this section, definitions are presented for a number of other key expressions which are generally used in the text and whose precise meaning must be made clear.

abnormal demand: demand records having given average and standard deviation values for each period of the year and day of the week, can be considered abnormal if they exceed, for the relevant period, values higher than (average + 2 × standard deviation).

Cases of abnormal demand are often combined with abrupt pressure drops in the system, as well as with increases in large leaks or pipe bursts.

active leakage control: a formal regime of regular (periodic) or continuous monitoring of the network that detects and *repairs unreported* bursts; includes regular surveying (sounding, waste metering) and/or leakage monitoring. Does not include *repair* of *reported* bursts.

Reported bursts are those that are notified to the relevant department of the *water utility*, either by own personnel or third parties, arising from visual evidence of leakage or found as a result of investigating customer complaint of loss of supply or low pressure.

advanced treatment: for the production of drinking water are the newly developed unit operations and processes – such as combination of UV and hydrogen peroxide – as well as those resulting from the refinement of conventional ones in order to meet more stringent quality criteria.

borehole water: water which is pumped out of an aquifer from below ground.

calibration: is a set of operations that establish, under specified conditions, the relationship between values of quantities indicated by a measuring instrument (or values represented by a material measure) and the corresponding values realised by standards.

The result of a calibration may be recorded in a document, e.g., a calibration certificate. The result can be expressed as corrections with respect to the indications of the instrument.

Calibration in itself does not necessarily mean that an instrument is performing in accordance with its specification.

cleaning: a range of techniques including flushing, swabbing, air scouring and pressure jetting intended to remove loose or soft deposits from inside water mains and storage tanks. Cleaning means the implementation of a formal procedure, as a rule written, the results of which are subsequently recorded.

complaints: any customer complaint expressed to the services either personally (verbal or written), by telephone, fax, post, e-mail or any other written format.

conventional treatment: for the production of drinking water includes the physical operations – such as activated carbon adsorption, filtration, membrane technologies – and the chemical processes – such as precipitation, disinfection, oxidation by ozone – that forms the basis of the line production in most water treatment plants.

delivery point: the point on the *service connection* where ownership of the pipe changes from the *water utility* to the customer (this is usually, but not always, at the boundary between the public highway and private land.)

effective meter reading: effective meter reading requires:

- the meter to be read accurately;
- the reading to be recorded correctly against the address of the customer;
- the customer bill to be generated using the meter reading;
- the bill to be delivered to the correct address;
- payment of the bill not to be delayed by any dispute over the meter reading.

inspection: implementation of a formal procedure, as a rule written, the results of which are subsequently recorded in a form that enables the *utility* to assess the serviceability of the assets and to take corrective action as appropriate.

integrated station: station with automatic control devices that are integrated in an automation network or are directly connected to a central control facility.

lowland surface water: a direct river abstraction, possibly with some storage. The upstream catchment may be well developed for agricultural and industrial uses.

long run marginal cost (LRMC): the summation of (a) the flow variable components of running cost (abstraction charges, energy, treatment chemicals and other process consumables), and (b) the incremental net present cost of providing additional capacity to meet rising demand and/or higher treated water quality standards, per cubic metre of water produced or saved.

Typical values for (a) would fall within the range 0.02 to 0.10 EUR/m^3. (b) would be nil where the *water utility* has available for the foreseeable future a sufficient supply of good quality water, but could as much as 0.50 EUR/m^3 or more if a new and expensive source would have to be developed to meet demand.

mains failure: detected water leaks of transmission and/or distribution mains necessitating *repair/renewal* measures. Included are failures of mains, defective pipe connections, valves and fittings, caused by:

- defective materials, design, construction or operational-related defects, in pipes, joints, valves and other fittings;
- corrosion of materials, externally or internally, primarily but not exclusively ferrous materials;
- external mechanical damage e.g., due to excavation, including third party damage;
- earth movements related to effects of frost, dry periods, heavy traffic, earthquakes, floods, and others.

Failures that occur in the joint between the main and the service connection shall be accounted for as mains failures if the failure requires interruption of the mains service and as service connection failure in the remaining situations.

Mains repairs from active leakage control shall not accounted for as mains failures[12].

marketing: includes a wide variety of functions such as advertising, mail-order business, public relations, retailing and merchandising, sales, wholesaling, market research, and pricing goods, related to the distribution and sale of goods or services. The aim of marketing is to acquire, retain, and satisfy customers.

[12] Repairs from active leakage control are in general related to non-apparent failures and their amount strongly depends on the scheduling of field campaigns. Therefore, their inclusion would cause bias and would prevent valid comparisons of failure rate indicators.

measurement point: the point on the *service connection* where customer meters are or would be located. In this definition, the customer meter may be either:

- a master meter serving two or more individual premises, which may or may not be individually metered; or
- an individual customer meter.

natural springs and wetlands water: water that naturally discharges from an aquifer at the surface owing to the interception of the water table with ground level.

rehabilitation: any physical intervention that extends the life of the system and/or improves its structural, hydraulic and/or water quality performance and involves changing its condition or specification.

In general, it refers to the system or sector and not to single pipes or other system elements. Structural rehabilitation includes *replacement* and *renovation*. Hydraulic rehabilitation includes *replacement*, reinforcement and, occasionally, *renovation*. Water quality rehabilitation includes *replacement* and *renovation*.

refurbishment: all methods for restoring the existing assets in order to achieve the requisite performance.

refurbishment is similar to *renovation* and is used mainly for pumps and other equipment.

reinforcement: construction of an additional facility that supplements the capacity of an existing one or provides an alternative to it.

relining: the removal of all deposits from inside an existing pipeline, followed by the *in situ* application of a non-structural lining to provide corrosion protection, such as cement or epoxy mortar (relining is sometimes referred to as scraping and lining, renovation or reconditioning).

Relining is a particular form of renovation.

renewal: is a particular form of *replacement* in which the function and characteristics of the new facility are the same as that of the existing.

In practice this usually means that it is of the same nominal diameter (for pipelines), power (for pumping systems), etc.

renovation: any physical intervention that extends the life of the asset and/or improves its performance, incorporating all or part of the existing asset, maintaining its initial function and capacity.

Renovation can include *repair*.

repair: rectification of local damage.

replacement: substitution of a new facility for an existing one where the latter is no longer used for its former objective.

The capacity of the new facility may or may not have the same capacity as the existing one. In the case of pipelines, replacement includes the provision of a structural liner (sliplining).

resident population: the population living in a given location on a permanent basis.

restrictions to water service: limitations to the use of water, as imposed by the *water utility*, such as hosepipe or sprinkler bans.

saline and brackish water: water that typically occurs at the margins of the freshwater system and the sea. May be found in a river estuary up to the tidal limit or in an aquifer, which underlies a coastal area. Saline water may also be found in borehole sources in inland salt bearing deposits.

service connection: the authorised pipe connecting the main to the measurement point or to the customer stop-valve, as applicable. Where several registered customers or individually occupied premises share a physical connection or tapping off the main, e.g., apartment buildings, this will still be regarded as the one connection for the purposes of the applicable PI, irrespective of the configuration and number of customers or premises. All active service connections shall be accounted for: connections to registered customers (residential and non-residential, temporary connections included), irrigation and fire hydrants, public taps or any other authorised consumption points not directly connected to the mains. Inactive connections to vacant buildings shall not be accounted for.

service connection failure: same definition as *mains failure* applied to *service connection.*

station: pumping stations, treatment plants and other control or monitoring points of the water supply system.

treated water: the water considered by the *utility* as fit for consumption, regardless of the treatment process it may or may not have been subject to. Treated water tests include all the tests carried out on treated water samples collected anywhere in the system, including the consumption points.

upland surface water: an impounding reservoir situated at relatively high altitude with little or no development in the upstream catchment.

water utility: whole set of organization, processes, activities, means and resources necessary for abstracting, treating, distributing or supplying drinking water or for collecting, treating and disposing of wastewater and for providing the associated services (definition 2.53 ISO 24512: 2007, including notes 1 to 4; in the context of this manual, the water utility services may include the provision of non-drinking water for urban uses)

NOTE 1 Some key features for a water utility are:

- its mission, to provide drinking water services or wastewater services, or both:
- its physical area of responsibility and the population within this area,
- its responsible body,
- the general organization with the function of operator being carried out by the responsible body, or by legally distinct operator(s),
- the type of physical systems used to provide the services, with various degrees of centralization.

NOTE 2 Drinking water utility addresses a utility dealing only with drinking water.

NOTE 3 When it is not necessary or it is difficult to make a distinction between responsible body and operator, the term "water utility" covers both.

NOTE 4 In common English, "water service" can be used as a synonym for "water utility", but the ISO 24510:2007 and ISO 24512: 2007 do not recommend using the term in this way.

well water: water which gravitates to an underground collection point through (usually) fissured strata, and which is then pumped out. Wells are man made, and may include horizontal excavations (adits), to extend the catchment to access additional fissures. Wells may also be sunk into the paths of underground streams. Another key example would be mine workings. Water quality would depend on the strata and overlying land use.

2. PERFORMANCE INDICATORS

2.1 Introduction

This chapter contains the detailed definition of the IWA performance indicators for water supply services. Each *utility* shall select exclusively the relevant subset for its specific case and objectives, according to the implementation procedure recommended in I-6.

If the *water utility* is responsible for other productive activities beyond water supply, only the core business component of the water supply shall be considered for the assessment of the indicators.

The IWA PI system aims to be used on an annual basis and therefore it is highly recommended that the year is used as the reference assessment period. However, since the *utilities* may need to track the evolution of their performance within the year, the PI system is prepared to accommodate other assessment periods for most indicators. In this case, and in order to ensure unit coherence and allow for PI comparison, all the PI expressed in terms of time are formulated in such a way that the values calculated for other assessment periods are converted into annual values.

Attention is drawn to the fact that the behaviour of most variables is not uniform during the year, due to random or seasonal effects, or to activity planning. All comparisons based on PI assessed from non-annual data should take this fact into consideration, in order to avoid bias.

Section II-2.2 contains a quick reference listing, and sections II-2.3 to II-2.8 the detailed information.

2.2 Quick reference listing of the IWA PI

Water resources indicators (pageII-202)	
WR1 –	Inefficiency of use of water resources (%)
WR2 –	Water resources availability (%)
	WR3 – Own water resources availability (%)
WR4 –	Reused supplied water (%)

Personnel indicators (page II-204)	
Total personnel (page II-204)	
Pe1 –	Employees per connection (No./1000 connections)
Pe2 –	Employees per water produced (No./(10^6 m^3/year))
Personnel per main function (page II-205)	
Pe3 –	General management personnel (%)
Pe4 –	Human resources management personnel (%)
Pe5 –	Financial and commercial personnel (%)
Pe6 –	Customer service personnel (%)
Pe7 –	Technical services personnel (%)
	Pe8 – Planning & construction personnel (%)
	Pe9 – Operations & maintenance personnel (%)
Technical services personnel per activity (page II-207)	
Pe10 –	Water resources and catchment management personnel (No./(10^6 m^3/year))
Pe11 –	Abstraction and treatment personnel (No./(10^6 m^3/year))
Pe12 –	Transmission, storage and distribution personnel (No./100 km)
Pe13 –	Water quality monitoring personnel (No./(10000 tests/year))
Pe14 –	Meter management personnel (No./1000 meters)
Pe15 –	Support services personnel (%)
Personnel qualification (page II-209)	
Pe16 –	University degree personnel (%)
Pe17 –	Basic education personnel (%)
Pe18 –	Other qualification personnel (%)
Personnel training (page II-210)	
Pe19 –	Total training (hours/employee/year)
	Pe20 – Internal training (hours/employee/year)
	Pe21 – External training (hours/employee/year)
Personnel health and safety (page II-211)	
Pe22 –	Working accidents (No./100 employees/year)
Pe23 –	Absenteeism (days/employee/year)
	Pe24 – Absenteeism due to working accidents or illness at work (days/employee/year)

	Pe25 – Absenteeism due to other reasons (days/employee/year)
Overtime work (page II-213)	
Pe26 –	Overtime work (%)

Physical indicators (page II-213)	
Treatment (page II-213)	
Ph1 –	Treatment plant utilisation (%)
Storage (page II-214)	
Ph2 –	Raw water storage capacity (days)
Ph3 –	Transmission and distribution storage capacity (days)
Pumping (page II-215)	
Ph4 –	Pumping utilisation (%)
Ph5 –	Standardised energy consumption (kWh/m^3/100 m)
Ph6 –	Reactive energy consumption (%)
Ph7 –	Energy recovery (%)
Transmission and distribution (page II-216)	
Ph8 –	Valve density (No./km)
Ph9 –	Hydrant density (No./km)
Meters (page II-217)	
	Ph10 – District meter density (No./1000 service connections)
	Ph11 – Customer meter density (No./service connection)
	Ph12 – Metered customers (No./customer)
	Ph13 – Metered residential customers (No./customer)
Automation and control (page II-218)	
Ph14 –	Automation degree (%)
Ph15 –	Remote control degree (%)

Operational indicators (page II-219)	
Inspection and maintenance of physical assets (page II-219)	
Op1 –	Pump inspection (-/year)
Op2 –	Storage tank cleaning (-/year)
Op3 –	Network inspection (%/year)

Op4 –	Leakage control (%/year)
Op5 –	Active leakage control repairs (No./100 km/year)
Op6 –	Hydrant inspection (-/year)
Instrumentation calibration (page II-222)	
Op7 –	System flow meters calibration (-/year)
Op8 –	Customer meter replacement (-/year)
Op9 –	Pressure meters calibration (-/year)
Op10 –	Water level meters calibration (-/year)
Op11 –	On-line water quality monitoring equipment calibration (-/year)
Electrical and signal transmission equipment inspection (page II-224)	
Op12 –	Emergency power system inspection (-/year)
Op13 –	Signal transmission equipment inspection (-/year)
Op14 –	Electrical switchgear equipment inspection (-/year)
Vehicle availability (page II-225)	
Op15 –	Vehicle availability (No./100 km)
Mains, valves and service connection rehabilitation (page II-226)	
Op16 –	Mains rehabilitation (%/year)
	Op17 – Mains renovation (%/year)
	Op18 – Mains replacement (%/year)
	Op19 – Replaced valves (%/year)
Op20 –	Service connection rehabilitation (%/year)
Pumps rehabilitation (page II-228)	
	Op21 – Pump refurbishment (%/year)
	Op22 – Pump replacement (%/year)
Operational water losses (page II-229)	
Op23 –	Water losses per connection (m^3/connection/year)
Op24 –	Water losses per mains length (m^3/km/day)
	Op25 – Apparent losses (%)
	Op26 – Apparent losses per system input volume (%)
	Op27 – Real losses per connection (l/connection/day when system is pressurised)
	Op28 – Real losses per mains length (l/km/day when system is pressurised)

Op29 –	Infrastructure leakage index (-)
Failure (page II-234)	
Op30 –	Pump failures (days/pump/year)
Op31 –	Mains failures (No./100 km/year)
Op32 –	Service connection failures (No./1000 connections/year)
Op33 –	Hydrant failures (No./1000 hydrants/year)
Op34 –	Power failures (hours/pumping station/year)
Water metering (page II-237)	
Op36 –	Customer reading efficiency (-)
Op37 –	Residential customer reading efficiency (-)
Op38 –	Operational meters (%)
Op39 –	Unmetered water (%)
Water quality monitoring (page II-239)	
Op40 –	Tests carried out (%)
	Op41 – Aesthetic tests carried out (%)
	Op42 – Microbiological tests carried out (%)
	Op43 – Physical–chemical tests carried out (%)
	Op44 – Radioactivity tests carried out (%)

Quality of service indicators (page II-241)	
Service coverage (page II-241)	
QS1 –	Households and businesses supply coverage (%)
QS2 –	Buildings supply coverage (%)
QS3 –	Population coverage (%)
	QS4 – Population coverage with service connections (%)
	QS5 – Population coverage with public taps or standpipes (%)
Public taps and standpipes (page II-244)	
QS6 –	Operational water points (%)
QS7 –	Average distance from waterpoints to households (m)
QS8 –	Per capita water consumed in public taps and standpipes (l/person/day)
QS9 –	Population per public tap or standpipe (persons/tap)

Pressure and continuity of supply (page II-245)	
QS10 –	Pressure of supply adequacy (%)
QS11 –	Bulk supply adequacy (%)
QS12 –	Continuity of supply (%)
QS13 –	Water interruptions (%)
QS14 –	Interruptions per connection (No./1000 connections/year)
QS15 –	Bulk supply interruptions (No./delivery point/year)
QS16 –	Population experiencing restrictions to water service (%)
QS17 –	Days with restrictions to water service (%)
Quality of supplied water (page II-249)	
QS18 –	Quality of supplied water (%)
	QS19 – Aesthetic tests compliance (%)
	QS20 – Microbiological tests compliance (%)
	QS21 – Physical-chemical tests compliance (%)
	QS22 – Radioactivity tests compliance (%)
Service connection and meter installation and repair (page II-251)	
QS23 –	New connection efficiency (days)
QS24 –	Time to install a customer meter (days)
QS25 –	Connection repair time (days)
Customer complaints (page II-252)	
QS26 –	Service complaints per connection (No. complaints/1000 connections/year)
QS27 –	Service complaints per customer (No. complaints/customer/year)
	QS28 – Pressure complaints (%)
	QS29 – Continuity complaints (%)
	QS30 – Water quality complaints (%)
	QS31 – Interruption complaints (%)
QS32 –	Billing complaints and queries (No./customer/year)
QS33 –	Other complaints and queries (No./customer/year)
QS34 –	Response to written complaints (%)

Economic and financial indicators (page II-256)	
Revenues (page II-256)	
Fi1 –	Unit revenue (EUR/m^3)
	Fi2 – Sales revenues (%)
	Fi3 – Other revenues (%)
Costs (page II-257)	
Fi4 –	Unit total costs (EUR/m^3)
	Fi5 – Unit running costs (EUR/m^3)
	Fi6 – Unit capital costs (EUR/m^3)
Composition of running costs per type of costs (page II-258)	
	Fi7 – Internal manpower costs (%)
	Fi8 – External services costs (%)
	Fi9 – Imported (raw and treated) water costs (%)
	Fi10 – Electrical energy costs (%)
	Fi11 – Other costs (%)
Composition of running costs per main function of the *water utility* (page II-260)	
	Fi12 – General management functions costs (%)
	Fi13 – Human resources management functions costs (%)
	Fi14 – Financial and commercial functions costs (%)
	Fi15 – Customer service functions costs (%)
	Fi16 – Technical services functions costs (%)
Composition of running costs per technical function activity (page II-261)	
	Fi17 – Water resources and catchment management costs (%)
	Fi18 – Abstraction and treatment costs (%)
	Fi19 – Transmission, storage and distribution costs (%)
	Fi20 – Water quality monitoring costs (%)
	Fi21 – Meter management costs (%)
	Fi22 – Support services costs (%)
Composition of capital costs (page II-263)	
	Fi23 – Depreciation costs (%)

	Fi24 – Net interest costs (%)
Investment (page II-264)	
Fi25 –	Unit investment (EUR/m^3)
	Fi26 – Investments for new assets and reinforcement of existing assets (%)
	Fi27 – Investments for asset replacement and renovation (%)
Average water charges (page II-266)	
Fi28 –	Average water charges for direct consumption (EUR/m^3)
Fi29 –	Average water charges for exported water (EUR/m^3)
Efficiency (page II-267)	
Fi30 –	Total cost coverage ratio (-)
Fi31 –	Operating cost coverage ratio (-)
Fi32 –	Delay in accounts receivable (days equivalent)
Fi33 –	Investment ratio (-)
Fi34 –	Contribution of internal sources to investment = CTI (%)
Fi35 –	Average age of tangible assets (%)
Fi36 –	Average depreciation ratio (-)
Fi37 –	Late payments ratio (-)
Fi38 –	Inventory value (-)
Leverage (page II-270)	
Fi39 –	Debt service coverage ratio = DSC (%)
Fi40 –	Debt equity ratio (-)
Liquidity (page II-270)	
Fi41 –	Current ratio (-)
Profitability (page II-271)	
Fi42	Return on net fixed assets (%)
Fi43	Return on equity (%)
Fi44	Return on capital employed (%)
Fi45	Asset turnover ratio (-)
Economic water losses (page II-272)	
Fi46 –	Non-revenue water by volume (%)
Fi47 –	Non-revenue water by cost (%)

2.3 Water resources indicators

WR1 – Inefficiency of use of water resources (%)
Real losses during the assessment period/system input volume during the assessment period × 100 **WR1 = A19/A3 × 100** A3 – System input volume (m^3) A19 – Real losses (m^3)
See also Op23, Op25, Op27 and Op29, and Fi46 and Fi47. It is recommended that this indicator is not assessed for periods shorter than one year, since it may lead to misleading conclusions. If a shorter assessment period cannot be avoided, special care is required in result interpretation. External comparisons on such time bases must be avoided. This indicator is not recommended to be used as a measure of efficiency of management of the transmission and/or the distribution system.

WR2 – Water resources availability (%)
(System input volume during the assessment period × 365/assessment period)/(annual yield capacity of own resources + annual imported water allowance) × 100 **WR2 = A3 × 365/H1/(A1 + A2) × 100** A1 – Annual yield capacity of own resources (m^3/year) A2 – Annual imported water allowance (m^3/year) A3 – System input volume (m^3) H1 – Assessment period (day)
It is recommended that this indicator is not assessed for periods shorter than one year, since it may lead to misleading conclusions. Annual values should ideally be analysed over a number of years and not in isolation. If a shorter assessment period cannot be avoided, special care is required in result interpretation. External comparisons on such time bases must be avoided. A value of 100% for this indicator means that all available resources are being used. Although this indicator is sometimes difficult to assess, and is not easily auditable, its used is encouraged as a management tool, particularly in rapid growing areas or areas subject to scarcity problems. Each *water utility* should estimate the annual abstraction capacity and imported water allowance taking into account water quality requirements and according to its own guaranteed schemes, drought management and operation procedures. This indicator is not suitable for comparisons, unless the same underlying assumptions are adopted to assess A1.

WR3 – Own water resources availability (%)

(System input volume during the assessment period × 365/assessment period)/ annual yield capacity of own resources × 100

WR3 = A3 × 365/H1/A1 × 100

A1 – Annual yield capacity of own resources (m^3/year)

A3 – System input volume (m^3)

H1 – Assessment period (day)

It is recommended that this indicator is not assessed for periods shorter than one year, since it may lead to misleading conclusions. Annual values should ideally be analysed over a number of years and not in isolation. If a shorter assessment period cannot be avoided, special care is required in result interpretation. External comparisons on such time bases must be avoided. This indicator is similar to WR2, but confined to the *utility* self-abstraction capacity. It is not applicable to *utilities* that count mostly on imported water.

WR4 – Reused supplied water (%)

Reused supplied water during the assessment period/system input volume during the assessment period × 100

WR4 = A22/A3 × 100

A3 – System input volume (m^3)

A22 – Reused supplied water (m^3)

It is recommended that this indicator is not assessed for periods shorter than one year, since it may lead to misleading conclusions. If a shorter assessment period cannot be avoided, special care is required in result interpretation. External comparisons on such time bases must be avoided. This indicator is applicable to systems when there is the need to use treated wastewater to complement the conventional available water resources.

2.4 Personnel indicators

Total personnel

Pe1 – Employees per connection (No./1000 connections)
Number of full time equivalent employees of the *water utility*/number of *service connections* × 1000 **Pe1 = B1/C24 × 1000** B1 – Total personnel (No.) C24 – Service connections (No.)
Indicators Pe1 and Pe2 shall be used in alternative. Pe2 applies if *service connections* density <20/km of mains (e.g., bulk supply systems). This indicator is assessed for a reference date.

Pe2 – Employees per water produced (No./(10^6 m^3/year))
Number of full time equivalent employees of the *water utility*/(water produced during the assessment period × 365/assessment period) × 10^6 **Pe2 = B1/(A6 × 365/H1) × 10^6** A6 – Water produced (m^3) B1 – Total personnel (No.) H1 – Assessment period (day)
Indicators Pe1 and Pe2 shall be used in alternative. Pe2 applies if *service connections* density <20/km of mains (e.g., bulk supply systems). This indicator may be assessed for periods shorter than one year, but special care is required in result interpretation when used for both internal or external comparisons.

Personnel per main function (see II-1.2 – Organisation functions for functions definition)

Pe3 – General management personnel (%)
Number of full time equivalent employees dedicated to directorate, central administration, strategic planning, marketing and comunications, other stakeholder relations, legal affairs, internal audits, environmental management, new business development and general co **Pe3 = B2/B1 × 100** B1 – Total personnel (No.) B2 – General management personnel (No.)
This indicator is assessed for a reference date.

Pe4 – Human resources management personnel (%)

Number of full time equivalent employees dedicated to personnel administration, education and training, occupational safety and health services and social activities/Number of full time equivalent employees of the *water utility* × 100

Pe4 = B3/B1 × 100

B1 – Total personnel (No.)

B3 – Human resources management personnel (No.)

This indicator is assessed for a reference date.

Pe5 – Financial and commercial personnel (%)

Number of full time equivalent employees dedicated to economic and financial planning, economic administration, economic controlling and purchasing and material management/Number of full time equivalent employees of the *water utility* × 100

Pe5 = B4/B1 × 100

B1 – Total personnel (No.)

B4 – Financial and commercial personnel (No.)

This indicator is assessed for a reference date.

Pe6 – Customer service personnel (%)

Number of full time equivalent employees dedicated to accounting and control and to customer relations and management activities/number of full time equivalent employees of the *water utility* × 100

Pe6 = B5/B1 × 100

B1 – Total personnel (No.)

B5 – Customer service personnel (No.)

This indicator is assessed for a reference date.

Pe7 – Technical services personnel (%)

Number of full time equivalent employees dedicated to planning, construction, operations and maintenance activities/number of full time equivalent employees of the *water utility* × 100

Pe7 = B6/B1 × 100

B1 – Total personnel (No.)

B6 – Technical services personnel (No.)

This indicator is assessed for a reference date.

PERFORMANCE INDICATORS

Pe8 – Planning & construction personnel (%)

Number of full time equivalent employees of technical services working in planning & construction/number of full time equivalent employees of the *water utility* × 100

Pe8 = B7/B1 × 100

B1 – Total personnel (No.)

B7 – Planning and construction personnel (No.)

This indicator is assessed for a reference date.

Pe9 – Operations & maintenance personnel (%)

Number of full time equivalent employees of technical services working in operations & maintenance/number of full time equivalent employees of the *water utility* × 100

Pe9 = B8/B1 × 100

B1 – Total personnel (No.)

B8 – Operations and maintenance personnel (No.)

This indicator is assessed for a reference date.

Technical services personnel per activity

(see II-1.2 – Organisation functions for functions definition)

Pe10 – Water resources and catchment management personnel (No./(10^6 m^3/year))

Number of full time equivalent employees working in the water resources and catchment management/(water produced during the assessment period × 365/assessment period) × 106

Pe10 = B9/(A6 × 365/H1) × 10^6

A6 – Water produced (m^3)

B9 – Water resources and catchment management personnel (No.)

H1 – Assessment period (day)

This indicator may be assessed for periods shorter than one year, but special care is required in result interpretation when used for both internal or external comparisons.

Pe11 – Abstraction and treatment personnel (No./(10⁶ m³/year))

Number of full time equivalent employees working in planning, design, construction, operations and maintenance of the abstraction and treatment/(water produced during the assessment period × 365/assessment period) × 106

Pe11 = B10/(A6 × 365/H1) × 10^6

A6 – Water produced (m^3)

B10 – Abstraction and treatment personnel (No.)

H1 – Assessment period (day)

This indicator may be assessed for periods shorter than one year, but special care is required in result interpretation when used for both internal or external comparisons.

Pe12 – Transmission, storage and distribution personnel (No./100 km)

Number of full time equivalent employees working in planning, design, construction, operations and maintenance of the transmission, storage and distribution system/total mains length × 100

Pe12 = B11/C8 × 100

B11 – Transmission, storage and distribution personnel (No.)

C8 – Mains length (km)

This indicator is assessed for a reference date.

Pe13 – Water quality monitoring personnel (No./(10000 tests/year))

Number of full time equivalent employees working in water quality sampling and testing/(total number of tests carried out by the *utility* laboratories during the assessment period × 365/assessment period) × 10000

Pe13 = B12/(D52 × 365/H1) × 10000

B12 – Water quality monitoring personnel (No.)

D52 – Water quality tests carried out (No.)

H1 – Assessment period (day)

This indicator may be assessed for periods shorter than one year, but special care is required in result interpretation when used for both internal or external comparisons.

Pe14 – Meter management personnel (No./1000 meters)
Number of full time equivalent employees working in meter management/total number of system and customer meters × 1000   **Pe14 = B13/(C10 + E6) × 1000**   B13 – Meter management personnel (No.)   C10 – System flow meters (No.)   E6 – Direct customer meters (No.)
This indicator is assessed for a reference date.

Pe15 – Support services personnel (%)
Number of full time equivalent employees working in support services/total number of full time equivalent employees working in technical services × 100   **Pe15 = B14/B6 × 100**   B6 – Technical services personnel (No.)   B14 – Support services personnel (No.)
This indicator is assessed for a reference date.

Personnel qualification

Pe16 – University degree personnel (%)
Number of full time equivalent employees of the *water utility* with university degree/ number of full time equivalent employees of the *water utility* × 100   **Pe16 = B15/B1 × 100**   B1 – Total personnel (No.)   B15 – University degree personnel (No.)
This indicator is assessed for a reference date.

Pe17 – Basic education personnel (%)
Number of other full time equivalent employees of the *water utility* with basic education/ number of full time equivalent employees of the *water utility* × 100   **Pe17 = B16/B1 × 100**   B1 – Total personnel (No.)   B16 – Basic education personnel (No.)
This indicator is assessed for a reference date

Pe18 – Other qualification personnel (%)
Number of full time equivalent employees of the *water utility* without basic education/ number of full time equivalent employees of the *water utility* × 100 **or** **100 – (university degree personnel + basic education personnel)** **Pe18 = B17/B1 × 100   or   Pe18 = 100 – Pe16 – Pe17** B1 – Total personnel (No.) B17 – Other qualification personnel (No.)
This indicator is assessed for a reference date.

Personnel training

Pe19– Total training (hours/employee/year)
(Number of training hours during the assessment period × 365/assessment period)/ number of full time equivalent employees of the *water utility* **Pe19 = (B18 × 365/H1)/B1** B1 – Total personnel (No.) B18 – Total training time (hour) H1 – Assessment period (day)
This indicator may be assessed for periods shorter than one year, but special care is required in result interpretation when used for both internal or external comparisons.

Pe20 – Internal training (hours/employee/year)
(Number of internal training hours/number of full time equivalent employees of the *water utility* **Pe20 = (B19 × 365/H1)/B1** B1 – Total personnel (No.) B19 – Internal training time (hour) H1 – Assessment period (day)
This indicator may be assessed for periods shorter than one year, but special care is required in result interpretation when used for both internal or external comparisons.

Pe21 – External training (hours/employee/year)

(Number of external training hours during the assessment period × 365/ assessment period)/number of full time equivalent employees of the *water utility*

Pe21 = (B20 × 365/H1)/B1

B1 – Total personnel (No.)

B20 – External training time (hour)

H1 – Assessment period (day)

This indicator may be assessed for periods shorter than one year, but special care is required in result interpretation when used for both internal or external comparisons

Personnel health and safety

Pe22 – Working accidents (No./100 employees/year)

(Number of working accidents requiring medical care occurring with personnel during the assessment period × 365/assessment period)/number of full time equivalent employees of the *water utility* × 100

Pe22 = (B21 × 365/H1)/B1 × 100

B1 – Total personnel (No.)

B21 – Working accidents (No.)

H1 – Assessment period (day)

This indicator may be assessed for periods shorter than one year, but special care is required in result interpretation when used for both internal or external comparisons.

Pe23 – Absenteeism (days/employee/year)

(Total number of days of absenteeism occurring during the assessment period × 365/ assessment period)/number of full time equivalent employees of the *water utility*

Pe23 = (B22 × 365/H1)/B1

B1 – Total personnel (No.)

B22 – Absenteeism (day)

H1 – Assessment period (day)

This indicator may be assessed for periods shorter than one year, but special care is required in result interpretation when used for both internal or external comparisons.

Pe24 – Absenteeism due to working accidents or illness at work (days/employee/year)

(Total number of days of absenteeism due to accidents or illness at work occurring during the assessment period × 365/assessment period)/number of full time equivalent employees of the *water utility*

Pe24 = (B23 × 365/H1)/B1

B1 – Total personnel (No.)

B23 – Absenteeism due to accidents or illness at work (days)

H1 – Assessment period (day)

This indicator may be assessed for periods shorter than one year, but special care is required in result interpretation when used for both internal or external comparisons.

Pe25 – Absenteeism due to other reasons (days/employee/year)

(Total number of days of absenteeism occurring during the assessment period that were not due to working accidents or illness at work × 365/assessment period)/number of full time equivalent employees of the *water utility*

Pe25 = (B24 × 365/H1)/B1

B1 – Total personnel (No.)

B24 – Absenteeism due to other reasons (day)

H1 – Assessment period (day)

This indicator may be assessed for periods shorter than one year, but special care is required in result interpretation when used for both internal or external comparisons.

Overtime work

Pe26 – Overtime work (%)

Overtime work during the assessment period/normal work during the assessment period × 100

Pe26 = B26/B25 × 100

B25 – Working time (hours)

B26 – Overtime work (hours)

This indicator may be assessed for periods shorter than one year, but special care is required in result interpretation when used for both internal or external comparisons. This indicator is meant for skilled and unskilled workers; too much overtime work indicates inefficient use of manpower, lack of personnel or inefficient preventive maintenance.

2.5 Physical indicators

Treatment

Ph1 – Treatment plant utilisation (%)
Maximum daily volume of water treated in treatment plants during the assessment period/ maximum daily capacity of the existing treatment plants × 100 **Ph1 = A4/C3 × 100** A4 – Maximum water treated daily (m^3/day) C3 – Daily treatment capacity (m^3/day)
It is recommended that this indicator is not assessed for periods shorter than one year, since it may lead to misleading conclusions. If a shorter assessment period cannot be avoided, special care is required in result interpretation. External comparisons on such time bases must be avoided.

Storage

Ph2 – Raw water storage capacity (days)
Net capacity of raw water reservoirs/system input volume during the assessment period × assessment period **Ph2 = C1/A3 × H1** A3 – System input volume (m^3) C1 – Raw water storage capacity (m^3) H1 – Assessment period (day)
It is recommended that this indicator is not assessed for periods shorter than one year, since it may lead to misleading conclusions. If a shorter assessment period cannot be avoided, special care is required in result interpretation. External comparisons on such time bases must be avoided.

Ph3 – Treated water storage capacity (days)

Total capacity of treated water reservoirs (private storage tanks excluded)/system input volume during the assessment period × assessment period

Ph3 = C2/A3 × H1

A3 – System input volume (m^3)

C2 – Treated water storage capacity (m^3)

H1 – Assessment period (day)

It is recommended that this indicator is not assessed for periods shorter than one year, since it may lead to misleading conclusions. If a shorter assessment period cannot be avoided, special care is required in result interpretation. External comparisons on such time bases must be avoided. In case of bulk supply systems, if the delivery point is a storage tank, its capacity can be accounted for, even though it is neither owned nor operator by the *water utility*. The interpretation of the values of this indicator shall take into account the seasonal, monthly, daily and hourly peak factors.

Pumping

Ph4 – Pumping utilisation (%)

Sum, for all installed pumps, of the number of operation hours of the maximum energy consumption day during the assessment period multiplied by the nominal power of the pump/(maximum nominal power that can be used simultaneously in the system × 24) × 10

Ph4 = D2/(C7 × 24) × 100

C7 – Maximum operating pumping capacity (kW)

D2 – Maximum daily pumping energy consumption (kWh)

This indicator measures the residual pumping capacity of the system for the day of overall maximum energy consumption. The denominator shall exclusively account for the installed pumps, including the backup ones, that can be used simultaneously in each pumping station.

Ph5 – Standardised energy consumption (kWh/m³/100 m)

Energy consumption for pumping during the assessment period/Sum of the volume elevated during the assessment period multiplied by the pump head/100)

Ph5 = D1/D3

D1 – Pumping energy consumption (kWh)

D3 – Standardisation factor ($m^3 \times 100$ m)

This indicator may be assessed for periods shorter than one year, but special care is required in result interpretation when used for both internal or external comparisons. This indicator is the average amount of energy consumed per m^3 at a pump head of 100 m. For reference, this indicator is normally of the order of 0,5 kWh/m^3 at 100 m. It is the inverse of the average pumping group efficiency. 0,5 kWh/m^3 at 100 m corresponds to an average pumping efficiency of 9810 N × 100 m/(3600 J/Wh)/500 Wh × 100 = 54%.

Ph6 – Reactive energy consumption (%)
Reactive energy consumption for pumping during the assessment period/total energy consumption for pumping during the assessment period × 100
Ph6 = D4/D1 × 100
D1 – Pumping energy consumption (kWh)
D4 – Reactive energy consumption (kVar)
This indicator may be assessed for periods shorter than one year, but special care is required in result interpretation when used for both internal or external comparisons. Reactive energy is normally expressed in kVAr and the energy consumption in kWh. However they are equivalent and their ratio is adimensional.

Ph7 – Energy recovery (%)
Energy recovered by the use of turbines of reverse pumps during the assessment period/ total energy consumption for pumping during the assessment period × 100
Ph7 = D5/D1 × 100
D1 – Pumping energy consumption (kWh)
D5 – Energy recovery (Wh)
This indicator may be assessed for periods shorter than one year, but special care is required in result interpretation when used for both internal or external comparisons.

Transmission and distribution

Ph8 – Valve density (No./km)
Number of isolating valves/total distribution mains length
Ph8 = C22/C9
C9 – Distribution mains length (km)
C22 – Isolating valves (No.)
This indicator is assessed for a reference date.

Ph9 – Hydrant density (No./km)
Number of hydrants/total distribution mains length
Ph9 = C23/C9
C9 – Distribution mains length (km)
C23 – Hydrants (No.)
This indicator is assessed for a reference date.

Meters

(all installed meters shall be accounted for, including those that are out of order; this comment is relevant for some developing regions where a significant part of installed meters are not working)

Ph10 – District meter density (No./1000 service connections)

Number of district meters/number of *service connections* × 1000

Ph10 = C11/C24 × 1000

C11 – District meters (No.)

C24 – Service connections (No.)

This indicator is assessed for a reference date.

Ph11 – Customer meter density (No./service connection)

Number of direct customer meters/number of *service connections*

Ph11 = E6/C24

C24 – Service connections (No.)

E6 – Direct customer meters (No.)

Ph12 – Metered customers (No./customer)

Number of direct and bulk customer meters/number of registered customers

Ph12 = (E6 + E9)/E10

E6 – Direct customer meters (No.)

E9 – Bulk customer meters (No.)

E10 – Registered customers (customer)

This indicator is assessed for a reference date.

Ph13 – Metered residential customers (No./customer)

Number of residential-equivalent customer meters/number of residential registered customers

Ph13 = E7/E11

E7 – Residential customer meters (No.)

E11 – Residential registered customers (customer)

This indicator is assessed for a reference date.

Automation and control

Ph14 – Automation degree (%)

Number of automated control units/number of control units × 100

Ph14 = C16/C15 × 100

C155 – Control units (No.)

C16 – Automated control units (No.)

This indicator is assessed for a reference date.

Ph15– Remote control degree (%)

Number of remotely controlled units/number of control units × 100

Ph15 = C17/C15 × 100

C15 – Control units (No.)

C17 – Remotely controlled units (No.)

This indicator is assessed for a reference date

2.6 Operational indicators

Inspection and maintenance of physical assets

Op1 – Pump inspection (-/year)

(Total nominal power of pumps and related ancillaries subjected to *inspection* during the assessment period × 365/assessment period)/total nominal power of pumps

Op1 = (D6 × 365/H1)/C6

C6 – Pumping stations capacity (kW)

D6 – Pumping inspection (kW)

H1 – Assessment period (day)

It is recommended that this indicator is not assessed for periods shorter than one year, since it may lead to misleading conclusions. If a shorter assessment period cannot be avoided, special care is required in result interpretation. External comparisons on such time bases must be avoided.

Op2 – Storage tank *cleaning* (-/year)

(Volume of storage tank cells cleaned during the assessment period × 365/assessment period)/total volume of storage tank cells

Op2 = (D7 × 365/H1)/C2

C2 – Transmission and distribution storage tanks capacity (m^3)

D7 – Storage tank cleaning (m^3)

H1 – Assessment period (day)

It is recommended that this indicator is not assessed for periods shorter than one year, since it may lead to misleading conclusions. Annual values should ideally be analysed over a number of years and not in isolation. If a shorter assessment period cannot be avoided, special care is required in result interpretation. External comparisons on such time bases must be avoided.

Op3 – Network inspection (%/year)

(Length of transmission and distribution mains where at least valves and other fittings were *inspected* during the assessment period × 365/assessment period)/total mains length × 100

Op3 = (D8 × 365/H1)/C8 × 100

C8 – Mains length (km)

D8 – Network inspection (km)

H1 – Assessment period (day)

It is recommended that this indicator is not assessed for periods shorter than one year, since it may lead to misleading conclusions. If a shorter assessment period cannot be avoided, special care is required in result interpretation. External comparisons on such time bases must be avoided.

Op4 – Leakage control (%/year)

(Length of mains subject to *active leakage control* during the assessment period × 365/assessment period)/total mains length × 100

Op4 = (D9 × 365/H1)/C8 × 100

C8 – Mains length (km)

D9 – Leakage control (km)

H1 – Assessment period (day)

It is recommended that this indicator is not assessed for periods shorter than one year, since it may lead to misleading conclusions. Annual values should ideally be analysed over a number of years and not in isolation. If a shorter assessment period cannot be avoided, special care is required in result interpretation. External comparisons on such time bases must be avoided.

Op5 – Active leakage control *repairs* (No./100 km/year)
(Number of leaks detected and repaired due to *active leakage control* during the assessment period × 365/assessment period)/total mains length × 100 **Op5 = (D10 × 365/H1)/C8 × 100** C8 – Mains length (km) D10 – Leaks repaired due to active leakage control (No.) H1 – Assessment period (day)
It is recommended that this indicator is not assessed for periods shorter than one year, since it may lead to misleading conclusions. Annual values should ideally be analysed over a number of years and not in isolation. If a shorter assessment period cannot be avoided, special care is required in result interpretation. External comparisons on such time bases must be avoided.

Op6 – Hydrant inspection (-/year)
(Number of hydrants *inspected* during the assessment period × 365/assessment period)/total number of hydrants **Op6 = (D11 × 365/H1)/C23** C23 – Hydrants (No.) D11 – Hydrant inspection (No.) H1 – Assessment period (day)
It is recommended that this indicator is not assessed for periods shorter than one year, since it may lead to misleading conclusions. If a shorter assessment period cannot be avoided, special care is required in result interpretation. External comparisons on such time bases must be avoided.

Instrumentation calibration

Op7 – System flow meters *calibration* (-/year)
(Number of system flowmeter *calibrations* carried out during the assessment period × 365/assessment period)/number of system flow meters installed in the system (permanently or temporarily) **Op7 = (D12 × 365/H1)/C10** C10 – System flow meters (No.) D12 – System flow meter calibrations (No.) H1 – Assessment period (day)
It is recommended that this indicator is not assessed for periods shorter than one year, since it may lead to misleading conclusions. If a shorter assessment period cannot be avoided, special care is required in result interpretation. External comparisons on such time bases must be avoided.

Op8 – Customer meter replacement (-/year)

(Number of customer flow meters replaced during the assessment period × 365/assessment period)/number of customer meters

Op8 = (D45 × 365/H1)/E6

D45 – Meter replacement (No.)

E6 – Direct customer meters (No.)

H1 – Assessment period (day)

It is recommended that this indicator is not assessed for periods shorter than one year, since it may lead to misleading conclusions. If a shorter assessment period cannot be avoided, special care is required in result interpretation. External comparisons on such time bases must be avoided.

Note: generally "calibration" of customer meters is held by replacing them by calibrated meters.

Op9 – Pressure meters *calibration* (-/year)

(Number of pressure meter *calibrations* carried out during the assessment period × 365/assessment period)/number of pressure meters installed in the system (permanently or temporarily)

Op9 = (D13 × 365/H1)/C12

C12 – Pressure meters (No.)

D13 – Pressure meter calibrations (No.)

H1 – Assessment period (day)

It is recommended that this indicator is not assessed for periods shorter than one year, since it may lead to misleading conclusions. If a shorter assessment period cannot be avoided, special care is required in result interpretation. External comparisons on such time bases must be avoided.

Op10 – Water level meters *calibration* (-/year)

(Number of water level meter *calibrations* carried out during the assessment period × 365/assessment period)/number of water level meters installed in the system (permanently or temporarily)

Op10 = (D14 × 365/H1)/C13

C13 – Water level meters (No.)

D14 – Water level meter calibrations (No.)

H1 – Assessment period (day)

It is recommended that this indicator is not assessed for periods shorter than one year, since it may lead to misleading conclusions. If a shorter assessment period cannot be avoided, special care is required in result interpretation. External comparisons on such time bases must be avoided.

Op11 – On-line water quality monitoring equipment *calibration* (-/year)
(Number of on-line water quality monitoring instrument *calibrations* carried out during the assessment period × 365/assessment period)/number of on-line water quality instruments installed in the system (permanently or temporarily) **Op11 = (D15 × 365/H1)/C14** C14 – On-line water quality monitoring instruments (No.) D15 – On-line water quality monitoring equipment calibrations (No.) H1 – Assessment period (day)
It is recommended that this indicator is not assessed for periods shorter than one year, since it may lead to misleading conclusions. If a shorter assessment period cannot be avoided, special care is required in result interpretation. External comparisons on such time bases must be avoided.

Electrical and signal transmission equipment inspection

Op12 – Emergency power system inspection (-/year)
(Sum of the nominal power of the emergency power systems *inspected* during the assessment period × 365/assessment period)/total nominal power of the emergency power systems **Op12 = (D16 × 365/H1)/C18** C18 – Emergency power systems (kW) D16 – Emergency power systems inspection (kW) H1 – Assessment period (day)
It is recommended that this indicator is not assessed for periods shorter than one year, since it may lead to misleading conclusions. Annual values should ideally be analysed over a number of years and not in isolation. If a shorter assessment period cannot be avoided, special care is required in result interpretation. External comparisons on such time bases must be avoided.

Op13 – Signal transmission equipment inspection (-/year)
(Number of signal transmission units *inspected* during the assessment period × 365/assessment period)/total number of signal transmission units **Op13 = (D17 × 365/H1)/C19** C19 – Signal transmission equipment (No.) D17 – Signal transmission equipment inspection (No.) H1 – Assessment period (day)
It is recommended that this indicator is not assessed for periods shorter than one year, since it may lead to misleading conclusions. If a shorter assessment period cannot be avoided, special care is required in result interpretation. External comparisons on such time bases must be avoided.

Op14 – Electrical switchgear equipment inspection (-/year)
(Number of electrical switchgear units *inspected* during the assessment period × 365/ assessment period)/total number of electrical switchgear units **Op14 = (D18 × 365/H1)/C20** C20 – Electrical switchgear (No.) D18 – Electrical switchgear inspection (No.) H1 – Assessment period (day)
It is recommended that this indicator is not assessed for periods shorter than one year, since it may lead to misleading conclusions. If a shorter assessment period cannot be avoided, special care is required in result interpretation. External comparisons on such time bases must be avoided.

Op15 – Vehicle availability (No./100 km)
Number of vehicles daily available, on a permanent basis, in average, for field works in operations and maintenance activities/total mains length × 100 **Op15 = D19/C8 × 100** C8 – Mains length (km) D19 – Permanent vehicles (No.)
It is recommended that this indicator is not assessed for periods shorter than one year, since it may lead to misleading conclusions. If a shorter assessment period cannot be avoided, special care is required in result interpretation. External comparisons on such time bases must be avoided. Heavy equipment shall not be included.

Mains, valves and service connection rehabilitation

Op16 – Mains *rehabilitation* (%/year)
(Length of transmission and distribution mains rehabilitated during the assessment period × 365/assessment period)/total mains length × 100 **Op16 = (D20 × 365/H1)/C8 × 100 or Op16 = Op17 + Op18** C8 – Mains length (km) D20 – Mains rehabilitation (km) H1 – Assessment period (day)
It is recommended that this indicator is not assessed for periods shorter than one year, since it may lead to misleading conclusions. Annual values should ideally be analysed over a number of years and not in isolation. If a shorter assessment period cannot be avoided, special care is required in result interpretation. External comparisons on such time bases must be avoided.

Op17 – Mains *renovation* (%/year)

(Length of mains renovated during the assessment period × 365/assessment period)/total mains length × 100

Op17 = (D21 × 365/H1)/C8 × 100

C8 – Mains length (km)

D21 – Mains renovation (km)

H1 – Assessment period (day)

It is recommended that this indicator is not assessed for periods shorter than one year, since it may lead to misleading conclusions. Annual values should ideally be analysed over a number of years and not in isolation. If a shorter assessment period cannot be avoided, special care is required in result interpretation. External comparisons on such time bases must be avoided.

Op18 – Mains *replacement* (%/year)

(Length of mains replaced during the assessment period × 365/assessment period)/total mains length × 100

Op18 = (D22 × 365/H1)/C8 × 100

C8 – Mains length (km)

D22 – Mains replacement (km)

H1 – Assessment period (day)

It is recommended that this indicator is not assessed for periods shorter than one year, since it may lead to misleading conclusions. Annual values should ideally be analysed over a number of years and not in isolation. If a shorter assessment period cannot be avoided, special care is required in result interpretation. External comparisons on such time bases must be avoided.

Op19 – Replaced valves (%/year)

(Number of mains valves replaced during the assessment period × 365/assessment period)/total number of mains valves × 100

Op19 = (D23 × 365/H1)/C21 × 100

C21 – Mains valves (No.)

D23 – Replaced valves (No.)

H1 – Assessment period (day)

It is recommended that this indicator is not assessed for periods shorter than one year, since it may lead to misleading conclusions. Annual values should ideally be analysed over a number of years and not in isolation. If a shorter assessment period cannot be avoided, special care is required in result interpretation. External comparisons on such time bases must be avoided.

Op20 – Service connection *rehabilitation* (%/year)

(Number of *service connections* replaced or renovated during the assessment period × 365/ assessment period)/total number of service connections × 100

Op20 = (D24 × 365/H1)/C24 × 100

C24 – Service connections (No.)

D24 – Service connection rehabilitation (No.)

H1 – Assessment period (day)

It is recommended that this indicator is not assessed for periods shorter than one year, since it may lead to misleading conclusions. Annual values should ideally be analysed over a number of years and not in isolation. If a shorter assessment period cannot be avoided, special care is required in result interpretation. External comparisons on such time bases m

Pumps rehabilitation

Op21 – Pump refurbishment (%/year)

(Total nominal power of pumps subject to overhaul during the assessment period × 365/assessment period)/total nominal power of pumps × 100

Op21 = (D25 × 365/H1)/C6 × 100

C6 – Pumping stations capacity (kW)

D25 – Pumps overhaul (kW)

H1 – Assessment period (day)

It is recommended that this indicator is not assessed for periods shorter than one year, since it may lead to misleading conclusions. Annual values should ideally be analysed over a number of years and not in isolation. If a shorter assessment period cannot be avoided, special care is required in result interpretation. External comparisons on such time bases must be avoided. Refurbishment may include partial replacement.

Op22 – Pump replacement (%/year)

(Total nominal power of pumps replaced during the assessment period × 365/ assessment period)/total nominal power of pumps × 100

Op22 = (D26 × 365/H1)/C6 × 100

C6 – Pumping stations capacity (kW)

D26 – Pumps replacement (kW)

H1 – Assessment period (day)

It is recommended that this indicator is not assessed for periods shorter than one year, since it may lead to misleading conclusions. Annual values should ideally be analysed over a number of years and not in isolation. If a shorter assessment period cannot be avoided, special care is required in result interpretation. External comparisons on such time bases must be avoided. Pump replacement refers exclusively to full replacement of the pump and the engine.

Water losses

See also WR1, and Fi46 and Fi47

Op23 – Water losses per connection (m^3/connection/year)

(Water losses during the assessment period × 365/assessment period)/number of *service connections*

Op23 = (A15 × 365/H1)/C24

A15 – Water losses (m^3)

C24 – Service connections (No.)

H1 – Assessment period (day)

Indicators Op23 and Op24 shall be used alternatively. Op24 applies if ***service connections*** density <20/km of mains (e.g., bulk supply systems).

It is recommended that this indicator is not assessed for periods shorter than one year, since it may lead to misleading conclusions. If a shorter assessment period cannot be avoided, special care is required in result interpretation. External comparisons on such time bases must be avoided.

Op24 – Water losses per mains length (m^3/km/year)

(Water losses during the assessment period/assessment period)/mains length

Op24 = (A15/H1)/C8

A15 – Water losses (m^3)

C8 – Mains length (km)

H1 – Assessment period (day)

Indicators Op23 and Op24 shall be used in alternative. Op24 applies if ***service connections*** density <20/km of mains (e.g., bulk supply systems).

It is recommended that this indicator is not assessed for periods shorter than one year, since it may lead to misleading conclusions. If a shorter assessment period cannot be avoided, special care is required in result interpretation. External comparisons on such time bases must be avoided.

Op25 – Apparent losses (%)

(Apparent losses/(system input volume – exported water) × 100

Op25 = A18/(A3 – A5 – A7) × 100

A18 – Apparent losses (m^3)

A3 – System input volume (m^3)

A5 – Exported raw water (m^3)

A7 – Exported treated water (m^3)

Indicators Op25 and Op26 shall be used in alternative. Op25 applies to distribution systems and Op26 to bulk supply systems.

For specific studies of apparent losses associated to customer meters, an indicator expressed as apparent losses per customer (m^3/customer) may be useful in alternative or in complement.

It is recommended that this indicator is not assessed for periods shorter than one year, since it may lead to misleading conclusions. If a shorter assessment period cannot be avoided, special care is required in result interpretation. External comparisons on such time bases must be avoided.

Op26 – Apparent losses per system input volume (%)

Apparent losses during the assessment period/system input volume × 100

Op26 = A18/A3

A18 – Apparent losses (m^3)

A3 – System input volume (m^3)

Indicators Op25 and Op26 shall be used in alternative. Op25 applies to distribution systems and Op26 to bulk supply systems.

It is recommended that this indicator is not assessed for periods shorter than one year, since it may lead to misleading conclusions. If a shorter assessment period cannot be avoided, special care is required in result interpretation. External comparisons on such time bases must be avoided.

Op27 – Real losses per connection (l/connection/day when system is pressurised)

Real losses during the assessment period × 1000/(number of *service connections* × number of hours system is pressurised during the assessment period/24)

Op27 = A19 × 1000/(C24 × H2/24/365)

A19 – Real losses (m^3)

C24 – Service connections (No.)

H2 – Time system is pressurised (hour)

Indicators Op27 and Op28 shall be used in alternative. Op28 applies if ***service connections*** density <20/km of mains (e.g., bulk supply systems).

It is recommended that this indicator is not assessed for periods shorter than one year, since it may lead to misleading conclusions. If a shorter assessment period cannot be avoided, special care is required in result interpretation. External comparisons on such time bases must be avoided.

Op28 – Real losses per mains length (l/km/day when system is pressurised)

Real losses during the assessment period × 1000/(mains length × number of hours system is pressurised during the assessment period/24)

Op28 = A19 × 1000/(C8 × H2/24)

A19 – Real losses (m^3)

C8 – Mains length (km)

H2 – Time system is pressurised (hour)

Indicators Op27 and Op28 shall be used in alternative. Op28 applies if ***service connections*** density <20/km of mains (e.g., bulk supply systems).

It is recommended that this indicator is not assessed for periods shorter than one year, since it may lead to misleading conclusions. If a shorter assessment period cannot be avoided, special care is required in result interpretation. External comparisons on such time bases must be avoided.

PERFORMANCE INDICATORS

Op29 – Infrastructure leakage index (-)

Real losses (Op27)/technical achievable low-level real losses (when system is pressurised)

Op29 = Op27/(18 × C8/C24 + 0,8 + 0,025 × C25)/(D34/10)

C8 – Mains length (km)

C24 – Service connections (No.)

C25 – Average service connection length (m)

D34 – Average operating pressure (kPa)

It is recommended that this indicator is not assessed for periods shorter than one year, since it may lead to misleading conclusions. If a shorter assessment period cannot be avoided, special care is required in result interpretation. External comparisons on such time bases must be avoided.

This indicator does not fulfill some of the requirements for a PI listed in I-2.2. However it is included in this manual because it is recommended by the IWA Water Losses Task Force.

The technical achievable low-level real losses are equal to the "best estimate" of so called Unavoidable Average Real Losses, UARL, which can be calculated with the equation derived by the Water Losses Task Force (see AQUA Dec. 1999 and IWA Blue Pages "Losses from water supply systems"):

UARL (litres/service connection/day)= (18 × Lm/Nc + 0.8 + 0.025 × Lp) × (P/10)

Corresponding to the following components:

On mains: 18 litres/km mains/day/metre of pressure

Plus, on sevice connections up to property boundary): 0.8 litres/service connection/day/metre of pressure

Plus, on service connections (property boundary to customer meter): 25 litres/km/day/metre of pressure

Variables used in this equation, on empirical results of international investigations, are:

– Length of mains Lm in km (C8)

– Number of *service connections* Nc (C24)

– Average length of *service connections* Lp in m (C25)

– Average operating pressure P in kPa (D34)

Well-managed systems are expected to have low values of this Infrastructure leakage index – close to 1.0 – while systems with infrastructure management deficiencies will present higher values.

Failures

Op30 – Pump failures (days/pump/year)

(Sum, for all pumps, of the number of days during the assessment period when the pump is out of order × 365/assessment period)/total number of pumps)

Op30 = (D27 × 365/H1)/C4

C4 – Pumps (No.)

D27 – Pump failures (day)

H1 – Assessment period (day)

It is recommended that this indicator is not assessed for periods shorter than one year, since it may lead to misleading conclusions. If a shorter assessment period cannot be avoided, special care is required in result interpretation. External comparisons on such time bases must be avoided.

Op31 – Mains failures (No./100 km/year)

(Number of *mains failures* during the assessment period (including failures of valves and fittings) × 365/assessment period)/total mains length × 100

Op31 = (D28 × 365/H1)/C8 × 100

C8 – Mains length (km)

D28 – Mains failures (No.)

H1 – Assessment period (day)

It is recommended that this indicator is not assessed for periods shorter than one year, since it may lead to misleading conclusions. If a shorter assessment period cannot be avoided, special care is required in result interpretation. External comparisons on such time bases must be avoided. If mains failures are to be used for regulating objectives, the use of a complementary indicator, similar to Op31 but excluding failures by third parties is advisable, as they are not a direct fault of the *water utility*. Numbers should exclude *repairs* under *active leakage control*.

Op32 – Service connection failures (No./1000 connections/year)

(Number of *service connection* failures during the assessment period × 365/assessment period)/number of *service connections* × 1000

Op32 = (D29 × 365/H1)/C24 × 1000

C24 – Service connections (No.)

D29 – Service connection failures (No.)

H1 – Assessment period (day)

It is recommended that this indicator is not assessed for periods shorter than one year, since it may lead to misleading conclusions. If a shorter assessment period cannot be avoided, special care is required in result interpretation. External comparisons on such time bases must be avoided. If *service connection* service connection failures are to be used for regulating objectives, the use of a complementary indicator, similar to *Op32* but excluding failures by third parties is advisable, as they are not a direct fault of the *water utility*. Numbers should exclude *repairs* under *active leakage control*.

Op33 – Hydrant failures (No./1000 hydrants/year)

(Number of hydrant failures during the assessment period × 365/assessment period)/total number of hydrants × 1000

Op33 = (D30 × 365/H1)/C23 × 1000

C23 – Hydrants (No.)

D30 – Hydrant failures (No.)

H1 – Assessment period (day)

It is recommended that this indicator is not assessed for periods shorter than one year, since it may lead to misleading conclusions. If a shorter assessment period cannot be avoided, special care is required in result interpretation. External comparisons on such time bases must be avoided. If hydrant failures are to be used for regulating objectives, the use of a complementary indicator, similar to Op33 but excluding failures by third parties is advisable, as they are not a direct fault of the *water utility*. Numbers should exclude *repairs* under *active leakage control*.

Op34 – Power failures (hours/pumping station/year)

Sum, for all pumps, of the number of hours each pumping station is out of service due to power supply interruptions during the assessment period × 365/assessment period)/total number of pumping stations

Op34 = (D31 × 365/H1)/C5

C5 – Pumping stations (No.)

D31 – Power failures (hour)

H1 – Assessment period (day)

It is recommended that this indicator is not assessed for periods shorter than one year, since it may lead to misleading conclusions. If a shorter assessment period cannot be avoided, special care is required in result interpretation. External comparisons on such time bases must be avoided. This indicator assesses the time that pumping stations are out of service due to power failures, regardless of the power source used (normal or standby power generation system). This means that the time when a pumping station is reliant on standby power generation shall not be considered as a power failure. This indicator aims to measure the performance of the water supply *utility* and not of the power supply *utility*.

Op35 – Water-point failures (No./water-point/year)

Number of water-points failures during the reference period/total number of waterpoints

Op35 = (D32 × 365/H1)/F6

D32 – Water-point failures (No.)

F6 – Water-points (No.)

H1 – Assessment period (day)

It is recommended that this indicator is not assessed for periods shorter than one year, since it may lead to misleading conclusions. If a shorter assessment period cannot be avoided, special care is required in result interpretation. External comparisons on such time bases must be avoided. This indicator is relevant for developing regions. Water point refers to public taps and standpipes. Water-point failures shall be accounted for regardless of their cause (tap, pump, piping, water source, etc.).

Water metering

Op36 – Customer reading efficiency (-)

(Number of *effective meter readings* effective meter readings carried out during the assessment period × 365/assessment period)/(number of residential customer meters × residential customer meter reading frequency + number of industrial customer meters × industrial customer meter reading frequency + number of bulk customer meters × bulk customer meter reading frequency)

Op36 = (D42 × 365/H1)/(E7 × D39 + E8 × D40 + E9 × D41)

D39 – Residential customer meter reading frequency (No./meter/year)

D40 – Industrial customer meter reading frequency (No./meter/year)

D41 – Bulk customer meter reading frequency (No./meter/year)

D42 – Customer meter readings (No.)

E7 – Residential customer meters (No.)

E8 – Industrial customer meters (No.)

E9 – Bulk customer meters (No.)

H1 – Assessment period (day)

Indicators Op36 and Op37 shall be used in alternative.

It is recommended that this indicator is not assessed for periods shorter than one year, since it may lead to misleading conclusions. If a shorter assessment period cannot be avoided, special care is required in result interpretation. External comparisons on such time bases must be avoided. This indicator expresses the percentage of expected meter readings that were effective. If there are other customer categories this indicator shall be adapted accordingly.

Op37 – Residential customer reading efficiency (-)

(Number of effective residential meter readings carried out during the assessment period × 365/assessment period)/(number of residential customer meters × residential customer meter reading frequency)

Op37 = (D43 × 365/H1)/(E7 × D39)

D39 – Residential customer meter reading frequency (No./meter/year)

D43 – Residential customer meter readings (No.)

E7 – Residential customer meters (No.)

H1 – Assessment period (day)

Indicators Op36 and Op37 shall be used in alternative.

It is recommended that this indicator is not assessed for periods shorter than one year, since it may lead to misleading conclusions. If a shorter assessment period cannot be avoided, special care is required in result interpretation. External comparisons on such time bases must be avoided. This indicator expresses the percentage of expected residential meter readings that were effective. Residential includes the set of customers that have the same procedures for meter reading as the metered residential consumers (i.e., commercial).

This indicator should only be used if Op36 cannot be calculated.

Op38 – Operational meters (%)

Number of direct customer meters installed that are not out-of-service at the reference time/number of direct customer meters × 100

Op38 = D44/E6 × 100

D44 – Operational meters (No.)

E6 – Direct customer meters (No.)

This indicator may be assessed for periods shorter than one year, but special care is required in result interpretation when used for both internal or external comparisons. This indicator is mainly relevant for developing regions where frequently a good number of installed meters are not operational.

Op39 – Unmetered water (%)

(System input volume- metered consumption)/system input volume during the assessment period × 100

Op39 = (A3 – A8 – A11)/A3 × 100

A3 – System input volume (m^3)

A8 – Billed metered consumption (m^3)

A11 – Unbilled metered consumption (m^3)

It is recommended that this indicator is not assessed for periods shorter than one year, since it may lead to misleading conclusions. If a shorter assessment period cannot be avoided, special care is required in result interpretation. External comparisons on such time bases must be avoided. Billed and unbilled metered consumption shall be included. It is recommended that this indicator is assessed on an annual basis only.

Water quality monitoring

These indicators aim to assess frequency compliance and therefore are limited to 100%. See variable definitions for clarity.

Op40 – Tests carried out (%)

Number of *treated water* tests carried out during the assessment period/number of *treated water* tests required by applicable standards or legislation during the assessment period × 100

Op40 = D46/D57 × 100

D46 – Required treated water quality tests carried out (No.)

D57 – Water quality tests required (No.)

This indicator may be assessed for periods shorter than one year, but special care is required in result interpretation when used for both internal or external comparisons.

Op41 – Aesthetic tests carried out (%)

Number of aesthetic tests of *treated water* carried out during the assessment period/number of aesthetic tests of *treated water* required by applicable standards or legislation during the assessment period × 100

Op41 = D47/D58 × 100

D47 – Required aesthetic tests carried out (No.)

D58 – Aesthetic tests required (No.)

This indicator may be assessed for periods shorter than one year, but special care is required in result interpretation when used for both internal or external comparisons.

Op42 – Microbiological tests carried out (%)

Number of microbiological tests of *treated water* carried out during the assessment period/number microbiological tests of *treated water* tests required by applicable standards or legislation during the assessment period × 100

Op42 = D48/D59 × 100

D48 – Required microbiological tests carried out (No.)

D59 – Microbiological tests required (No.)

This indicator may be assessed for periods shorter than one year, but special care is required in result interpretation when used for both internal or external comparisons.

Op43 – Physical-chemical tests carried out (%)

Number of physical-chemical tests of *treated water* carried out during the assessment period/number of physical-chemical tests of *treated water* required by applicable standards or legislation during the assessment period × 100

Op43 = D49/D60 × 100

D49 – Required physical-chemical tests carried out (No.)

D60 – Physical-chemical tests required (No.)

This indicator may be assessed for periods shorter than one year, but special care is required in result interpretation when used for both internal or external comparisons.

Op44 – Radioactivity tests carried out (%)

Number of radioactivity tests of *treated water* carried out during the assessment period/number of radioactivity tests of *treated water* required by applicable standards or legislation during the assessment period × 100

Op44 = D50/D61 × 100

D50 – Required radioactivity tests carried out (No.)

D61 – Radioactivity tests required (No.)

This indicator may be assessed for periods shorter than one year, but special care is required in result interpretation when used for both internal or external comparisons.

2.7 Quality of service indicators

Coverage

QS1 – Households and businesses supply coverage (%)

Number of households and businesses connected to the public network/total number of households and businesses × 100

QS1 = E1/E3 × 100

E1 – Households and businesses supplied (No.)

E3 – Households and businesses (No.)

Indicators QS1, QS2 and QS3 shall be used in alternative.

This indicator is assessed for a reference date. This indicator, QS1, shall be adopted to assess service coverage in countries where there is a registered customer per household and there is reliable statistical information about the number of households and businesses. In such cases, this indicator is a more reliable and advisable alternative (particularly in areas with season-dependent population) to the population coverage indicator (QS3) that is currently more often adopted.

QS2 – Buildings supply coverage (%)

Number of buildings connected to the public network/total number of buildings × 100

QS2 = E2/E4 × 100

E2 – Buildings supplied (No.)

E4 – Buildings (No.)

Indicators QS1, QS2 and QS3 shall be used in alternative.

This indicator is assessed for a reference date. In case QS1 is not applicable, the building supply coverage indicator (QS2) is the most advisable indicator to assess service coverage, particularly in areas with season-dependent population.

QS3 – Population coverage (%)

***resident population* served by the *water utility*/total *resident population* × 100**

QS3 = F1/E5 × 100

E5 – Resident population (person)

F1 – Population supplied (person)

Indicators QS1, QS2 and QS3 shall be used in alternative.

This indicator is assessed for a reference date. QS3 is currently the indicator of service coverage more widely used. However, as it makes use of data that most water *utilities* cannot assess accurately (population served), it shall be used only when QS1 and QS2 are not applicable.

QS4 – Population coverage with *service connections* (%)

***resident population* served by the *water utility* through *service connections*/total *resident population* × 100**

QS4 = F2/E5 × 100

E5 – Resident population (person)

F2 – Population supplied with service pipes (person)

This indicator is assessed for a reference date. This indicator is relevant only for developing regions where a significant part of the supply is provided by public taps and standpipes.

QS5 – Population coverage with public taps or standpipes (%)

***Resident population* served by the *water utility* through public taps or standpipes/ total *resident population* × 100**

QS5 = F3/E5 × 100

E5 – Resident population (person)

F3 – Population served by public taps or standpipes (person)

This indicator is assessed for a reference date. This indicator is relevant only for developing regions where a significant part of the supply is provided by public taps and standpipes.

Public taps and standpipes
(when applicable)

QS6 – Operational water points (%)

Number of water points that are not out-of-service/number of water points × 100

QS6 = F7/F6 × 100

F6 – Water-points (No.)

F7 – Operational water-points (No.)

This indicator is assessed for a reference date. Water points refer to consumption points where one or more public taps or standpipes are located. A water point shall be considered out-of-service if it cannot supply any water to the consumers due to asset failures (e.g., broken taps, out-of-order pumps). This indicator is more relevant for rural and peri-urban areas of developing regions.

QS7 – Average distance from waterpoints to households (m)

Sum, for all water points, of the distance between the water point and the far-most household served by it/total number of water points

QS7 = F4/F6

F4 – Distance from water points to households (m)

F6 – Water-points (No.)

This indicator is assessed for a reference date. Water points refer to consumption points where one or more public taps or standpipes are located. This indicator is relevant only for developing regions where a significant part of the supply is provided by public taps and standpipes.

QS8 – Per capita water consumed in public taps and standpipes (l/person/day)

(Sum, for all water points, of the water consumption at the water point during the assessment period × 1000)/*resident population* served by the *water utility* through public taps or standpipes/assessment period

QS8 = F5 × 1000/F3/H1

F3 – Population served by public taps or standpipes (person)

F5 – Public taps and standpipes consumption (m^3/year)

H1 – Assessment period (day)

It is recommended that this indicator is not assessed for periods shorter than one year, since it may lead to misleading conclusions. If a shorter assessment period cannot be avoided, special care is required in result interpretation. External comparisons on such time bases must be avoided. This indicator is relevant only for developing regions where a significant part of the supply is provided by public taps and standpipes. In some cases the assessment of this indicator may need to be based on estimates.

QS9 – Population per public tap or standpipe (persons/tap)
Resident population **served by the** ***water utility*** **through public taps or standpipes/number of public taps and standpipes**
QS9 = F3/F8
F3 – Population served by public taps or standpipes (person)
F8 – Public taps and standpipes (No.)
This indicator is assessed for a reference date. This indicator is relevant only for developing regions where a significant part of the supply is provided by public taps and standpipes.

Pressure and continuity of supply

QS10 – Pressure of supply adequacy (%)
Number of ***delivery points*** **that receive and are likely to receive pressure equal to or above the guaranteed or declared target level at the peak demand hour (but not when demand is** ***abnormal*****)/number of** ***service connections*** **× 100**
QS10 = D33/C24 × 100
C24 – Service connections (No.)
D33 – Delivery points with adequate pressure (No.)
Indicators QS10 and QS11 shall be used as alternatives. QS11 applies if *service connections* density <20/km of mains (e.g., bulk supply systems).
It is recommended that this indicator is not assessed for periods shorter than one year, since it may lead to misleading conclusions. If a shorter assessment period cannot be avoided, special care is required in result interpretation. External comparisons on such time bases must be avoided.

QS11 – Bulk supply adequacy (%)
Number of ***delivery points*** **that are supplied at any time according to the target flow, volume and/or pressure/number of** ***delivery points*** **× 100**
QS11 = D33/E9 × 100
D33 – Delivery points with adequate pressure (No.)
E9 – Bulk customer meters (No.)
Indicators QS10 and QS11 shall be used as alternatives. QS11 applies if *service connections* density <20/km of mains (e.g., bulk supply systems).
It is recommended that this indicator is not assessed for periods shorter than one year, since it may lead to misleading conclusions. If a shorter assessment period cannot be avoided, special care is required in result interpretation. External comparisons on such time bases must be avoided. Target flow, volume and/or pressure can be contracted or agreed between the supplier and the bulk customers. The processing rule of this indicator assumes that every *delivery point* is metered. If this is not the case, the user needs to create a new variable "number of *delivery points*".

QS12 – Continuity of supply (%)

Number of hours when the system is pressurised during the assessment period/24/ assessment period × 100

QS12 = H2/24/H1 × 100

H1 – Assessment period (day)

H2 – Time system is pressurised (hour)

It is recommended that this indicator is not assessed for periods shorter than one year, since it may lead to misleading conclusions. If a shorter assessment period cannot be avoided, special care is required in result interpretation. External comparisons on such time bases must be avoided. This indicator is relevant for intermittent supply systems. Interruptions due to unplanned system failures or to on-going *repair* or rehabilitation works shall not be accounted for by this indicator. When there are subsystems supplied in different periods, the indicator has to be assessed individually for each subsystem and the result is a weighted average using the number of *service connections* of each subsystem as weighting factor.

QS13 – Water interruptions (%)

Sum, for the assessment period, of the population subject to a water multiplied by the respective duration of the interruption in hours/(population served × 24 × assessment period) × 100

QS13 = D35/(F1 × 24 × H1) × 100

D35 – Water interruptions (person × hour)

F1 – Population supplied (person)

H1 – Assessment period (day)

Indicators QS13, QS14 and QS15 shall be used as alternatives, the former being more informative but also more complex to assess. Indicator QS14 may be applied when *service connections* density ≥20/km of mains (urban distribution systems); QS15 may be applied when *service connections* density <20/km of mains (rural distribution systems and bulk supply systems).

It is recommended that this indicator is not assessed for periods shorter than one year, since it may lead to misleading conclusions. If a shorter assessment period cannot be avoided, special care is required in result interpretation. External comparisons on such time bases must be avoided. Since, for many water *utilities*, the information required for this indicator is neither available nor feasible to be collected in a near future, QS14 or QS15 are alternatively proposed.

QS14 – Interruptions per connection (No./1000 connections/year)

(Total number of interruptions during the assessment period × 365/assessment period)/ number of *service connections* × 1000

QS14 = (D36 × 365/H1)/C24 × 1000

C24 – Service connections (No.)

D36 – Service interruptions (No.)

H1 – Assessment period (day)

Indicators QS13, QS14 and QS15 shall be used as alternatives, the former being more informative but also more complex to assess. Indicator QS14 may be applied when *service connections* density ≥20/km of mains (urban distribution systems); QS15 may be applied when *service connections* density <20/km of mains (rural distribution systems and bulk supply systems).

It is recommended that this indicator is not assessed for periods shorter than one year, since it may lead to misleading conclusions. If a shorter assessment period cannot be avoided, special care is required in result interpretation. External comparisons on such time bases must be avoided.

This indicator should only be used if QS13 cannot be calculated.

QS15 – Bulk supply interruptions (No./*delivery point*/year)

(Total number of interruptions during the assessment period × 365/assessment period)/ number of *delivery points*

QS15 = (D36 × 365/H1)/E9

D36 – Service interruptions (No.)

E9 – Bulk customer meters (No.)

H1 – Assessment period (day)

Indicators QS13, QS14 and QS15 shall be used as alternatives, the former being more informative but also more complex to assess. Indicator QS14 may be applied when *service connections* density ≥20/km of mains (urban distribution systems); QS15 may be applied when *service connections* density <20/km of mains (rural distribution systems and bulk supply systems).

It is recommended that this indicator is not assessed for periods shorter than one year, since it may lead to misleading conclusions. If a shorter assessment period cannot be avoided, special care is required in result interpretation. External comparisons on such time bases must be avoided. The processing rule of this indicator assumes that every *delivery point* delivery point is metered. If this is not the case, the user needs to create a new variable "number of *delivery points*".

This indicator should only be used if QS13 cannot be calculated.

QS16 – Population experiencing restrictions to water service (%)

Sum, for the assessment period, of the population affected by *restrictions to water service* multiplied by the respective duration of the *restrictions to water service* in hours/(total population served × 24 × assessment period) × 100

QS16 = D37/(F1 × 24 × H1) × 100

D37 – Water use restrictions (person × hour)

F1 – Population supplied (person)

H1 – Assessment period (day)

Indicators QS16 and QS17 shall be used as alternatives.

It is recommended that this indicator is not assessed for periods shorter than one year, since it may lead to misleading conclusions. If a shorter assessment period cannot be avoided, special care is required in result interpretation. External comparisons on such time bases must be avoided.

QS17 – Days with restrictions to water service (%)

Total number of days with *restrictions to water service* during the assessment period/ assessment period × 100

QS17 = D38/H1 × 100

D38 – Days with restrictions to water service (day)

H1 – Assessment period (day)

Indicators QS16 and QS17 shall be used as alternatives.

It is recommended that this indicator is not assessed for periods shorter than one year, since it may lead to misleading conclusions. Annual values should ideally be analysed over a number of years and not in isolation. If a shorter assessment period cannot be avoided, special care is required in result interpretation. External comparisons on such time bases must be avoided. This indicator should only be used if QS16 cannot be calculated.

Quality of supplied water

QS18 – Quality of supplied water (%)

Total number of *treated water* tests complying with the applicable standards or legislation during the assessment period/total number of tests of *treated water* carried out during the assessment period × 100

QS18 = (D62 + D63 + D64 + D65)/D51 × 100

D51 – Treated water quality tests carried out (No.)

D62 – Compliance of aesthetic tests (No.)

D63 – Compliance of microbiological tests (No.)

D64 – Compliance of physical-chemical tests (No.)

D65 – Compliance of radioactivity tests (No.)

This indicator may be assessed for periods shorter than one year, but special care is required in result interpretation when used for both internal or external comparisons.

QS19 – Aesthetic tests compliance (%)

Number of aesthetic tests of treated water complying with the applicable standards or legislation during the assessment period/total number of aesthetic tests of *treated water* carried out during the assessment period × 100

QS19 = D62/D53 × 100

D53 – Aesthetic tests carried out (No.)

D62 – Compliance of aesthetic tests (No.)

This indicator may be assessed for periods shorter than one year, but special care is required in result interpretation when used for both internal or external comparisons.

QS20 – Microbiological tests compliance (%)

Number of microbiological tests of *treated water* complying with the applicable standards or legislation during the assessment period/total number of microbiological tests of *treated water* carried out during the assessment period × 100

QS20 = D63/D54 × 100

D54 – Microbiological tests carried out (No.)

D63 – Compliance of microbiological tests (No.)

This indicator may be assessed for periods shorter than one year, but special care is required in result interpretation when used for both internal or external comparisons.

QS21 – Physical-chemical tests compliance (%)

Number of physical-chemical tests of *treated water* complying with the applicable standards or legislation during the assessment period/total number of physical-chemical tests of *treated water* carried out during the assessment period × 100

QS21 = D64/D55 × 100

D55 – Physical-chemical tests carried out (No.)

D64 – Compliance of physical-chemical tests (No.)

This indicator may be assessed for periods shorter than one year, but special care is required in result interpretation when used for both internal or external comparisons.

QS22 – Radioactivity tests compliance (%)

Number of radioactivity tests of *treated water* complying with the applicable standards or legislation during the assessment period/total number of radioactivity tests of *treated water* carried out during the assessment period × 100

QS22 = D65/D56 × 100

D56 – Radioactivity tests carried out (No.)

D65 – Compliance of radioactivity tests (No.)

This indicator may be assessed for periods shorter than one year, but special care is required in result interpretation when used for both internal or external comparisons.

Service connection and meter installation and repair

QS23 – New connection efficiency (days)

Total time spent for establishing new connections during the assessment period/number of new connections installed during the assessment period

QS23 = F9/F10

F9 – New connections establishment time (day)

F10 – New connections established (No.)

It is recommended that this indicator is not assessed for periods shorter than one year, since it may lead to misleading conclusions. If a shorter assessment period cannot be avoided, special care is required in result interpretation. External comparisons on such time bases must be avoided. This indicator refers to new contracts, when *service connections* already exist.

QS24 – Time to install a customer meter (days)
Total time spent for installing customer meters during the assessment period/number of customer meters installed during the assessment period **QS24 = F11/F12** F11 – Customer meter installation time (day) F12 – New customer meters installed (No.)
It is recommended that this indicator is not assessed for periods shorter than one year, since it may lead to misleading conclusions. If a shorter assessment period cannot be avoided, special care is required in result interpretation. External comparisons on such time bases must be avoided. This indicator refers exclusively to existing connections that have been newly fitted with a meter.

QS25 – Connection *repair* time (days)
Total time spent for repairing service connections during the assessment period/total number of connections repaired during the assessment period **QS25 = F13/F14** F13 – Connections repair time (day) F14 – Connections repaired (No.)
It is recommended that this indicator is not assessed for periods shorter than one year, since it may lead to misleading conclusions. If a shorter assessment period cannot be avoided, special care is required in result interpretation. External comparisons on such time bases must be avoided.

Customer complaints

QS26 – Service complaints per connection (No. complaints/1000 connections/year)
(Number of *complaints* of quality of service during the assessment period × 365/assessment period)/number of *service connections* × 1000 **QS26 = (F15 × 365/H1)/C24 × 1000** C24 – Service connections (No.) F15 – Service complaints (No.) H1 – Assessment period (day)
Indicators QS26 and QS27 shall be used as alternatives. QS27 applies if *service connections* density <20/km of mains (e.g., bulk supply systems). This indicator may be assessed for periods shorter than one year, but special care is required in result interpretation when used for both internal or external comparisons.

QS27 – Service complaints per customer (No. complaints/customer/year)

(Number of *complaints* of quality of service during the assessment period × 365/assessment period)/number of bulk supply customers

QS27 = (F15 × 365/H1)/E10

E10 – Registered customers (customer)

F15 – Service complaints (No.)

H1 – Assessment period (day)

Indicators QS26 and QS27 shall be used as alternatives. QS27 applies if *service connections* density <20/km of mains (e.g., bulk supply systems).

This indicator may be assessed for periods shorter than one year, but special care is required in result interpretation when used for both internal or external comparisons.

QS28 – Pressure complaints (%)

Number of pressure *complaints* during the assessment period/number of service *complaints* during the assessment period × 100

QS28 = F16/F15 × 100

F15 – Service complaints (No.)

F16 – Pressure complaints (No.)

This indicator may be assessed for periods shorter than one year, but special care is required in result interpretation when used for both internal or external comparisons.

QS29 – Continuity complaints (%)

Number of continuity *complaints* during the assessment period/number of service *complaints* during the assessment period × 100

QS29 = F17/F15 × 100

F15 – Service complaints (No.)

F17 – Continuity complaints (No.)

This indicator may be assessed for periods shorter than one year, but special care is required in result interpretation when used for both internal or external comparisons. This indicator refers to medium- or long-term supply constraints, due to insufficient water quantity or quality at source, insufficient system capacity or on-going works.

QS30 – Water quality complaints (%)

Number of water quality *complaints* during the assessment period/number of service *complaints* during the assessment period × 100

QS30 = F18/F15 × 100

F15 – Service complaints (No.)

F18 – Water quality complaints (No.)

This indicator may be assessed for periods shorter than one year, but special care is required in result interpretation when used for both internal or external comparisons.

QS31 – Interruption complaints (%)

Number of *complaints* due to supply interruptions during the assessment period/ number of service *complaints* during the assessment period × 100

QS31 = F19/F15 × 100

F15 – Service complaints (No.)

F19 – Complaints on interruptions (No.)

This indicator may be assessed for periods shorter than one year, but special care is required in result interpretation when used for both internal or external comparisons. This indicator refers to short-term supply interruptions, due to accidental system failures or *repair* works.

QS32 – Billing complaints and queries (No./customer/year)

(Number of billing *complaints* and queries during the assessment period × 365/assessment period)/number of registered customers

QS32 = (F20 × 365/H1)/E10

E10 – Registered customers (customer)

F20 – Billing complaints and queries (No.)

H1 – Assessment period (day)

QS33 – Other complaints & queries (No./customer/year)

(Number of other *complaints* and queries during the assessment period × 365/assessment period)/number of registered customers

QS33 = (F21 × 365/H1)/E10

E10 – Registered customers (customer)

F21 – Other complaints and queries (No.)

H1 – Assessment period (day)

This indicator may be assessed for periods shorter than one year, but special care is required in result interpretation when used for both internal or external comparisons.

QS34 – Response to written complaints (%)
Number of written responses within the target time during the assessment period/number of written complaints during the assessment period × 100 **QS34 = F22/F23 × 100** F22 – Written responses (No.) F23 – Written complaints (No.)
This indicator may be assessed for periods shorter than one year, but special care is required in result interpretation when used for both internal or external comparisons. This indicator is applicable in case there is a Guaranteed Standards Scheme.

2.8 Economic and financial indicators

Revenues

Fi1 – Unit revenue (EUR/m³)
(Operating revenues – capitalised costs of self constructed assets)/authorised consumption (including exported water), during the assessment period **Fi1 = (G2 – G35)/A14** A14 – Authorised consumption (m³) G2 – Operating revenues (EUR) G35 – Capitalised cost of self-constructed assets (EUR)
Note that G2–G35 = G1 (total revenues). It is recommended that this indicator is not assessed for periods shorter than one year, since it may lead to misleading conclusions. If a shorter assessment period cannot be avoided, special care is required in result interpretation. External comparisons on such time bases must be avoided. See section 1.3, Table 68 and Table 69 for definitions.

Fi2 – Sales revenues (%)
(Sales revenues/total revenues) × 100, during the assessment period **Fi2 = G3/G1 × 100** G1 – Total revenues (EUR) G3 – Sales revenues (EUR)
It is recommended that this indicator is not assessed for periods shorter than one year, since it may lead to misleading conclusions. If a shorter assessment period cannot be avoided, special care is required in result interpretation. External comparisons on such time bases must be avoided. See section 1.3, Table 68 and Table 69 for definitions.

Fi3 – Other revenues (%)
Other revenues not coming from sales/total revenues × 100, during the assessment period
Fi3 = (G1 – G3)/G1 × 100
G1 – Total revenues (EUR)
G3 – Sales revenues (EUR)
It is recommended that this indicator is not assessed for periods shorter than one year, since it may lead to misleading conclusions. If a shorter assessment period cannot be avoided, special care is required in result interpretation. External comparisons on such time bases must be avoided. See section 1.3, Table 68 and Table 69 for definitions.

Costs

(See section 1.3 – Financial definitions)

Fi4 – Unit total costs (EUR/m^3)
(Running costs + capital costs)/authorised consumption (including exported water), during the assessment period
Fi4 = G4/A14
A14 – Authorised consumption (m^3)
G4 – Total costs (EUR)
It is recommended that this indicator is not assessed for periods shorter than one year, since it may lead to misleading conclusions. If a shorter assessment period cannot be avoided, special care is required in result interpretation. External comparisons on such time bases must be avoided. See section 1.3, Table 68 and Table 69 for definitions.

Fi5 – Unit running costs (EUR/m^3)
Running costs/authorised consumption (including exported water), during the assessment period
Fi5 = G5/A14
A14 – Authorised consumption (m^3)
G5 – Running costs (EUR)
This indicator may be assessed for periods shorter than one year, but special care is required in result interpretation when used for both internal or external comparisons. See section 1.3, Table 68 and Table 69 for definitions.

Fi6 – Unit capital costs (EUR/m^3)
Capital costs/authorised consumption (including exported water), during the assessment period
Fi6 = G6/A14
A14 – Authorised consumption (m^3)
G6 – Capital costs (EUR)
It is recommended that this indicator is not assessed for periods shorter than one year, since it may lead to misleading conclusions. If a shorter assessment period cannot be avoided, special care is required in result interpretation. External comparisons on such time bases must be avoided. See section 1.3, Table 68 and Table 69 for definitions.

Composition of running costs per type of costs

Fi7 – Internal manpower costs (%)
Internal manpower costs/running costs × 100, during the assessment period
Fi7 = G8/G5 × 100
G5 – Running costs (EUR)
G8 – Internal manpower costs (EUR)
This indicator may be assessed for periods shorter than one year, but special care is required in result interpretation when used for both internal or external comparisons.

Fi8 – External services costs (%)
External services costs/running costs × 100, during the assessment period
Fi8 = G9/G5 × 100
G5 – Running costs (EUR)
G9 – External services costs (EUR)
This indicator may be assessed for periods shorter than one year, but special care is required in result interpretation when used for both internal or external comparisons.

Fi9 – Imported (raw and treated) water costs (%)

Imported (raw and treated) water costs/running costs, during the assessment period × 100

Fi9 = G10/G5 × 100

G5 – Running costs (EUR)

G10 – Imported (raw and treated) water costs (EUR)

This indicator may be assessed for periods shorter than one year, but special care is required in result interpretation when used for both internal or external comparisons.

Fi10 – Electrical energy costs (%)

Electrical energy costs/running costs × 100, during the assessment period

Fi10 = G11/G5 × 100

G5 – Running costs (EUR)

G11 – Electrical energy costs (EUR)

This indicator may be assessed for periods shorter than one year, but special care is required in result interpretation when used for both internal or external comparisons.

Fi11 – Other costs (%)

(Purchased merchandises + leasing and rentals + taxes, levies & fees + exceptional earnings & losses + other operating costs)/running costs × 100, during the assessment period

Fi11 = (G12 + G13 + G14 + G15 + G16)/G5 × 100

G5 – Running costs (EUR)

G12 – Purchased merchandises (EUR)

G13 – Leasing and rentals (EUR)

G14 – Taxes, levies and fees (EUR)

G15 – Exceptional earnings and losses (EUR)

G16 – Other operating costs (EUR)

This indicator may be assessed for periods shorter than one year, but special care is required in result interpretation when used for both internal or external comparisons.

Composition of running costs per main function of the *water utility*

Fi12 – General management functions costs (%)
Running costs of general management functions/running costs × 100, during the assessment period **Fi12 = G17/G5 × 100** G5 – Running costs (EUR) G17 – General management running costs (EUR)
This indicator may be assessed for periods shorter than one year, but special care is required in result interpretation when used for both internal or external comparisons. See section 1.2, Table 62 and section 1.3, Table 69 for definitions.

Fi13 – Human resources management functions costs (%)
Running costs of human resources management functions/running costs × 100, during the assessment period **Fi13 = G18/G5 × 100** G5 – Running costs (EUR) G18 – Human resources management running costs (EUR)
This indicator may be assessed for periods shorter than one year, but special care is required in result interpretation when used for both internal or external comparisons. See section 1.2, Table 63 and section 1.3, Table 69 for definitions.

Fi14 – Financial and commercial functions costs (%)
Running costs of financial and commercial functions/running costs × 100, during the assessment period **Fi14 = G19/G5 × 100** G5 – Running costs (EUR) G19 – Financial and commercial running costs (EUR)
This indicator may be assessed for periods shorter than one year, but special care is required in result interpretation when used for both internal or external comparisons. See section 1.2, Table 64, and section 1.3, Table 69 for definitions.

Fi15 – Customer service functions costs (%)
Running costs of customer service functions/running costs × 100, during the assessment period **Fi15 = G20/G5 × 100** G5 – Running costs (EUR) G20 – Customer service running costs (EUR)
This indicator may be assessed for periods shorter than one year, but special care is required in result interpretation when used for both internal or external comparisons. See section 1.2, Table 65, and section 1.3, Table 69 for definitions.

Fi16 – Technical services functions costs (%)
Running costs of the technical functions: planning, design, construction, operations and maintenance/running costs × 100, during the assessment period **Fi16 = G21/G5 × 100** G5 – Running costs (EUR) G21 – Planning, design, construction, operations & maintenance running costs (EUR)
This indicator may be assessed for periods shorter than one year, but special care is required in result interpretation when used for both internal or external comparisons. See section 1.2, Table 66 and Table 67 and section 1.3, Table 69 for definitions.

Composition of running costs per technical function activity
(see 1.2 – Organisation functions for functions definition)

Fi17 – Water resources and catchment management costs (%)
Running costs of the water resources and catchment management/running costs × 100, during the assessment period **Fi17 = G22/G5 × 100** G5 – Running costs (EUR) G22 – Water resources and catchment management running costs (EUR)
This indicator may be assessed for periods shorter than one year, but special care is required in result interpretation when used for both internal or external comparisons.

Fi18 – Abstraction and treatment costs (%)

Running costs of the abstraction and treatment/running costs × 100, during the assessment period

Fi18 = G23/G5 × 100

G5 – Running costs (EUR)

G23 – Abstraction and treatment running costs (EUR)

This indicator may be assessed for periods shorter than one year, but special care is required in result interpretation when used for both internal or external comparisons.

Fi19 – Transmission, storage and distribution costs (%)

Running costs of the transmission, storage and distribution/running costs × 100, during the assessment period

Fi19 = G24/G5 × 100

G5 – Running costs (EUR)

G24 – Transmission, storage and distribution running costs (EUR)

This indicator may be assessed for periods shorter than one year, but special care is required in result interpretation when used for both internal or external comparisons.

Fi20 – Water quality monitoring costs (%)

Running costs of the water quality sampling and testing/running costs × 100, during the assessment period

Fi20 = G25/G5 × 100

G5 – Running costs (EUR)

G25 – Water quality sampling and testing running costs (EUR)

This indicator may be assessed for periods shorter than one year, but special care is required in result interpretation when used for both internal or external comparisons.

Fi21 – Meter management costs (%)

Running costs of the meter management/running costs × 100, during the assessment period

Fi21 = G26/G5 × 100

G5 – Running costs (EUR)

G26 – Meter management running costs (EUR)

This indicator may be assessed for periods shorter than one year, but special care is required in result interpretation when used for both internal or external comparisons.

Fi22 – Support services costs (%)

Running costs of the support services/running costs × 100, during the assessment period

Fi22 = G27/G5 × 100

G5 – Running costs (EUR)

G27 – Support services running costs (EUR)

This indicator may be assessed for periods shorter than one year, but special care is required in result interpretation when used for both internal or external comparisons.

Composition of capital costs

Fi23 – Depreciation costs (%)

Depreciation costs/capital costs × 100, during the assessment period

Fi23 = G28/G6 × 100

G6 – Capital costs (EUR)

G28 – Depreciation costs (EUR)

It is recommended that this indicator is not assessed for periods shorter than one year, since it may lead to misleading conclusions. If a shorter assessment period cannot be avoided, special care is required in result interpretation. External comparisons on such time bases must be avoided.

Fi24 – Net interest costs (%)

(Interest expenses costs – interest income)/capital costs × 100, during the assessment period

Fi24 = (G29 – G30)/G6 × 100

G6 – Capital costs (EUR)

G29 – Interest expenses costs (EUR)

G30 – Interest income (EUR)

It is recommended that this indicator is not assessed for periods shorter than one year, since it may lead to misleading conclusions. If a shorter assessment period cannot be avoided, special care is required in result interpretation. External comparisons on such time bases must be avoided.

Investments

Fi25 – Unit investment (EUR/m³)

Cost of investments (expenditures for plant and equipment)/authorised consumption (including exported water), during the assessment period

Fi25 = G32/A14

A14 – Authorised consumption (m^3)

G32 – Investment in water service assets (EUR)

It is recommended that this indicator is not assessed for periods shorter than one year, since it may lead to misleading conclusions. Annual values should ideally be analysed over a number of years and not in isolation. If a shorter assessment period cannot be avoided, special care is required in result interpretation. External comparisons on such time bases must be avoided.

Fi26 – Investments for new assets and *reinforcement* of existing assets (%)

Cost of investments for new assets and *reinforcement* of existing assets/total cost of the investments × 100, during the assessment period

Fi26 = G33/G32 × 100

G32 – Investment in water service assets (EUR)

G33 – Investments for new assets and reinforcement of existing assets (EUR)

It is recommended that this indicator is not assessed for periods shorter than one year, since it may lead to misleading conclusions. Annual values should ideally be analysed over a number of years and not in isolation. If a shorter assessment period cannot be avoided, special care is required in result interpretation. External comparisons on such time bases must be avoided.

Fi27 – Investments for asset *replacement* and *renovation* (%)

Cost of investments for the *replacement* and *renovation* ("like for like") of existing assets/total cost of the investments × 100, during the assessment period

Fi27 = G34/G32 × 100

G32 – Investment in water service assets (EUR)

G34 – Investments for asset replacement and renovation (EUR)

"Like for like" means that it will provide the same functionality. It is recommended that this indicator is not assessed for periods shorter than one year, since it may lead to misleading conclusions. Annual values should ideally be analysed over a number of years and not in isolation. If a shorter assessment period cannot be avoided, special care is required in result interpretation. External comparisons on such time bases must be avoided.

Average water charges
(without public taxes)

Fi28 – Average water charges for direct consumption (EUR/m^3)

Water sales revenue from residential, industrial and other customers (exported water excluded; public water taxes excluded)/(total authorised- exported water), during the assessment period

Fi28 = G36/(A14 – A7)

A7 – Exported treated water (m^3)

A14 – Authorised consumption (m^3)

G36 – Water sales revenue for direct consumption (EUR)

It is recommended that this indicator is not assessed for periods shorter than one year, since it may lead to misleading conclusions. If a shorter assessment period cannot be avoided, special care is required in result interpretation. External comparisons on such time bases must be avoided.

Fi29 – Average water charges for exported water (EUR/m^3)

Water sales revenue from exported water (excluding public water taxes)/exported water, during the assessment period

Fi29 = G37/(A5 + A7)

A5 – Exported raw water (m^3)

A7 – Exported treated water (m^3)

G37 – Water sales revenue for exported water (EUR)

It is recommended that this indicator is not assessed for periods shorter than one year, since it may lead to misleading conclusions. If a shorter assessment period cannot be avoided, special care is required in result interpretation. External comparisons on such time bases must be avoided.

Efficiency indicators

Fi30 – Total cost coverage ratio (-)

Total revenues/total costs, during the assessment period

Fi30 = G1/G4

G1 – Total revenues (EUR)

G4 – Total costs (EUR)

It is recommended that this indicator is not assessed for periods shorter than one year, since it may lead to misleading conclusions. If a shorter assessment period cannot be avoided, special care is required in result interpretation. External comparisons on such time bases must be avoided. See section II-1.2, Table 68 and Table 69 for definitions.

PERFORMANCE INDICATORS

Fi31 – Operating cost coverage ratio (-)

Total revenues/running costs, during the assessment period

Fi31 = G1/G5

G1 – Total revenues (EUR)

G5 – Running costs (EUR)

It is recommended that this indicator is not assessed for periods shorter than one year, since it may lead to misleading conclusions. If a shorter assessment period cannot be avoided, special care is required in result interpretation. External comparisons on such time bases must be avoided. See section II-3, Table 68 and Table 69 for definitions.

Fi32 – Delay in accounts receivable (days equivalent)

Accounts receivable from drinking water at the reference date/sales revenues during the assessment period × assessment period

Fi32 = G38/G3 × H1

G3 – Sales revenues (EUR)

G38 – Accounts receivable (EUR)

It is recommended that this indicator is not assessed for periods shorter than one year, since it may lead to misleading conclusions. If a shorter assessment period cannot be avoided, special care is required in result interpretation. External comparisons on such time bases must be avoided.

Fi33 – Investment ratio (-)

Investments subject to depreciation/depreciation costs, during the assessment period

Fi33 = G39/G28

G28 – Depreciation costs (EUR)

G39 – Investments subject to depreciation (EUR)

Investments subject to depreciation are included in the investments reported in section II-1.3, Table 71. It is recommended that this indicator is not assessed for periods shorter than one year, since it may lead to misleading conclusions. Annual values should ideally be analysed over a number of years and not in isolation. If a shorter assessment period cannot be avoided, special care is required in result interpretation. External comparisons on such time bases must be avoided.

Fi34 – Contribution of internal sources to investment = CTI (%)
Investments financed by the cash flow/total investments × 100, during the assessment period **Fi34 = G40/G32 × 100** G32 – Investment in water service assets (EUR) G40 – Investments financed by the cash flow (EUR)
It is recommended that this indicator is not assessed for periods shorter than one year, since it may lead to misleading conclusions. If a shorter assessment period cannot be avoided, special care is required in result interpretation. External comparisons on such time bases must be avoided. See section II-1.3, Table 71 for cash flow definition.

Fi35 – Average age of water service assets (%)
Depreciated historical value of water service assets/historical value of water service assets × 100, during the year **Fi35 = G41/G42 × 100** G41 – Depreciated historical value of water service assets (EUR/year) G42 – Historical value of water service assets (EUR/year)
This indicator must be assessed on an annual basis. If in the balance sheet the water service values are not re-evaluated with inflation, section II-1.3, Table 72 shows the net value of water service assets as the difference between the historical and the depreciated values. This indicator expresses the average age of water service assets in terms of passed technical and economic life given 100 the total.

Fi36 – Average depreciation ratio (-)
Depreciation costs/historical value of water service assets, during the year **Fi36 = G28/G42** G28 – Depreciation costs (EUR) G42 – Historical value of water service assets (EUR/year)
This indicator must be assessed on an annual basis. See section II-1.3, Table 68 and Table 69 for annual depreciation definition.

Fi37 – Late payments ratio (-)
[1 – (annual debt from customers/amount billed during the year)], during the year
Fi37 = [1 – (G43/G44)]
G43 – Annual debt from customers (EUR/year)
G44 – Annual amount billed per year (EUR/year)
This indicator must be assessed on an annual basis.

Fi38 – Inventory value (-)
Value of overall inventory at the end of the fiscal year/operating revenues during the year
Fi38 = G51/G2
G2 – Operating revenues (EUR)
G51 – Inventories (EUR)
This indicator must be assessed on an annual basis.

Leverage indicators

Fi39 – Debt service coverage ratio = DSC (%)
Cash-flow/financial debt service × 100, during the assessment period
Fi39 = G45/G46 × 100
G45 – Cash-flow (EUR)
G46 – Financial debt service (EUR)
It is recommended that this indicator is not assessed for periods shorter than one year, since it may lead to misleading conclusions. If a shorter assessment period cannot be avoided, special care is required in result interpretation. External comparisons on such time bases must be avoided. See section II-1.3, Table 71 for cash flow definition. The financial debt service contains the expenditures for interest, the cost of loans, and the principal repayment debt instruments.

Fi40 – Debt equity ratio (-)
Total debt/shareholders' equity, at the end of the fiscal year
Fi40 = G47/G48
G47 – Total debt (EUR)
G48 – Shareholders' equity (EUR)
This indicator must be assessed on an annual basis. Water *utilities* with high debt service coverage ratio tend to have low debt equity ratio. It is also affected by revaluation of fixed assets and therefore caution should be exercised when using this ratio.

Liquidity indicator

Fi41 – Current ratio (-)
Current assets/current liabilities, at the reference date **Fi41 = G49/G53** G49 – Current assets (EUR) G53 – Current liabilities (EUR)
This indicator is assessed for a reference date. See section II-1.3, Table 68, Table 69 and Table 72 for definitions. This ratio measures the short-run paying ability of the *water utility*.

Profitability indicators

Fi42 – Return on net fixed assets (%)
Operating income/(historical value of water service assets – depreciated historical value of water service assets) × 100, during the year **Fi42 = G54/(G42 – G41) × 100** G41 – Depreciated historical value of water service assets (EUR/year) G42 – Historical value of water service assets (EUR/year) G54 – Operating income (EUR/year)
It measures the productivity of fixed assets in use. This indicator shall be assessed on an annual basis. Annual values should ideally be analysed over a number of years and not in isolation. This indicator measures the return on fixed assets (ROA), relating net income to the net value of water service fixed assets. See section II-1.3, Table 68, Table 69 and Table 72 for definitions.

Fi43 – Return on equity (%)
Net income (net income after interest payment and taxes)/shareholders' equity × 100, during the year **Fi43 = G56/G48 × 100** G48 – Shareholders' equity (EUR) G56 – Net income (EUR/year)
This indicator must be assessed on an annual basis. Annual values should ideally be analysed over a number of years and not in isolation. This indicator measures the return on equity (ROE), relating net income to the shareholders equity. See section II-1.3, Table 68, Table 69 and Table 72 for definitions.

Fi44 – Return on capital employed (%)
(Operating income (EBIT) – related taxes)/total assets × 100, during the year **Fi44 = (G54 – G55)/G50 × 100** G50 – Total assets (EUR) G54 – Operating income (EUR/year) G55 – Operating income related taxes (EUR/year)
This indicator must be assessed on an annual basis. See section II-1.3, Table 68, Table 69 and Table 72 for definitions. This indicator (RoCE) expresses the net return that the shareholder gains from his employed capital.

Fi45 – Asset turnover ratio (-)
Sales revenues/total assets, during the year **Fi45 = G3/G50** G3 – Sales revenues (EUR) G50 – Total assets (EUR)
It is recommended that this indicator is not assessed for periods shorter than one year, since it may lead to misleading conclusions. If a shorter assessment period cannot be avoided, special care is required in result interpretation. External comparisons on such time bases must be avoided. See section II-1.3, Table 68, Table 69 and Table 72 for definitions.

Water losses indicators

See also WR1, and Op23, Op25, Op27 and Op29.

Fi46 – Non-revenue water by volume (%)
Non-revenue water/system input volume, during the assessment period × 100 **Fi46 = A21/A3 × 100** A3 – System input volume (m^3) A21 – Non-revenue water (m^3)
It is recommended that this indicator is not assessed for periods shorter than one year, since it may lead to misleading conclusions. If a shorter assessment period cannot be avoided, special care is required in result interpretation. External comparisons on such time bases must be avoided.

Fi47 – Non-revenue water by cost (%)
Valuation of non-revenue water components/running costs, during the assessment period × 100 **Fi47 = ((A13 + A18) × G57 + A19 × G58)/G5 × 100** A13 – Unbilled authorised consumption (m^3) A18 – Apparent losses (m^3) A19 – Real losses (m^3) G5 – Running costs (EUR) G57 – Average water charges for direct consumption (EUR/m^3) G58 – Attributed unit cost for real losses (EUR/m^3)
It is recommended that this indicator is not assessed for periods shorter than one year, since it may lead to misleading conclusions. If a shorter assessment period cannot be avoided, special care is required in result interpretation. External comparisons on such time bases must be avoided. This is the sum of separate valuations for unbilled authorised consumption, apparent losses and real losses. It is usually worthwhile to calculate and review the three components of Fi47 separately.

3. VARIABLES

This section contains the specification of each variable required to derive the performance indicators proposed. It has the following objectives:

- to allow for a comprehensive interpretation of the concepts described in the body text;
- to serve as a guide for the creation of an information system for performance indicators assessment;
- to support audit trails.

Each *utility* shall assess exclusively the variables required by the performance indicators previously selected.

If the *water utility* is responsible for other productive activities beyond water supply, only the core business component of the water supply shall be considered for the assessment of the variables.

Each variable form contains the same type of information. For variables assessed at a given time (e.g., number of employees), it is highly advisable that the same date is adopted for all the variables requiring a reference date. A similar rule applies to the variables that refer to a pre-set period of assessment (e.g., pipe failures).

It is most advisable that the target confidence grades are defined for every input data variable, according to I-6 and II-5.

3.1 Section A – Water volume data

A1 – Annual yield capacity of own resources (m³/year)
Maximum annual volume of water that can potentially be abstracted from own resources, based on the availability of water resources and on any legal or contractual allowance constraints. INPUT DATA Referred to a reference period
If the maximum annual yield capacity is not clearly established as an allowance, it shall be estimated as accurately as possible based on technical studies. In the latter case, the assessment of this variable requires an hydrological study that takes into account the definition of failure scenarios resulting for instance from scarcity or water quality problems, their probability of risk and general water resources management procedures. Constraints derived from the abstraction infrastructures shall not be considered. USED FOR INDICATORS: WR2, WR3

A2 – Annual imported water allowance (m³/year)
Maximum allowance of raw and treated water importation. INPUT DATA Referred to a reference period
If the maximum annual allowance is not contracted with the supplier, an estimate based on the knowledge of the existing situation may be used. If no data to support this estimate exists, the sum of the imported raw water and of the imported treated water shall be used. USED FOR INDICATORS: WR2

Table 79. Components of water balance specifying the variable codes.

<table>
<tr><td rowspan="10">System input volume A3</td><td rowspan="4">Authorised consumption A14 = A10 + A13</td><td rowspan="2">Billed authorised consumption A10 = A8 + A9</td><td>Billed metered consumption A8</td><td rowspan="2">Revenue water A20 = A8 + A9</td></tr>
<tr><td>Billed non-metered consumption A9</td></tr>
<tr><td rowspan="2">Unbilled authorised consumption A13 = A11 + A12</td><td>Unbilled metered consumption A11</td><td rowspan="8">Non-revenue water A21 = A3 – A20</td></tr>
<tr><td>Unbilled non-metered consumption A12</td></tr>
<tr><td rowspan="6">Water losses A15 = A3 – A14</td><td rowspan="2">Apparent losses A18 = A16 + A17</td><td>Unauthorised consumption A16</td></tr>
<tr><td>Metering inaccuracies water losses A17</td></tr>
<tr><td rowspan="4">Real losses A19 = A15 – A18</td><td>Real losses on raw water mains and at the treatment works (if appropriate)</td></tr>
<tr><td>Leakage on transmission and/or distribution mains</td></tr>
<tr><td>Leakage and overflows at transmission and/or distribution storage tanks</td></tr>
<tr><td>Leakage on service connections up to the measurement point</td></tr>
</table>

A3 – System input volume (m^3)

The water volume input of the global system during the assessment period.

INPUT DATA

Referred to a reference period

System input should include water abstracted and all imported water (raw and treated).

System input value may be a sum of metered and unmetered volumes, corrected with the corresponding systematic error estimates.

See section 1.1 for definitions and Table 79 – Components of water balance specifying the variable codes.

USED FOR INDICATORS:	WR1, WR2, WR3, WR4, Ph2, Ph3, Op23(A15), Op24(A15), Op25, Op26, Op27(A19), Op28(A19), Op39, Fi46, Fi47(A19)
USED FOR VARIABLES:	A15, A19(A15), A21
ALSO USED FOR CI:	CI70

A4 – Maximum water treated daily (m^3/day)

Maximum daily volume of water treated in treatment plants during the assessment period.

INPUT DATA

Referred to a reference period

This variable is the maximum of the sum of the individual daily volumes treated in the system, and not the sum of the individual daily maximums treated by each treatment plant, in order to take into account that the treatment peaks are not simultaneous in all treatment plants.

USED FOR INDICATORS: Ph1

A5 – Exported raw water (m^3)

Total volume of raw water transferred to other *water utility* or to another system from the same supply area during the assessment period.

INPUT DATA

Referred to a reference period

See definitions of section II-1.1 – Water balance definitions.

USED FOR INDICATORS:	Op25, Fi29
ALSO USED FOR CI:	CI69, CI70, CI72

A6 – Water produced (m^3)

Total volume of water treated for input to water transmission lines or directly to the distribution system during the assessment period.

INPUT DATA

Referred to a reference period

Imported treated water should not be included. See section II-1.1 for definitions.

USED FOR INDICATORS:	Pe2, Pe10, Pe11

A7 – Exported treated water (m^3)

Total volume of treated water exported to other *water utility* or to another system from the same supply area during the assessment period.

INPUT DATA

Referred to a reference period

These transfers can occur anywhere downstream of the treatment plants or at any point where the water is assumed as treated by the *water utility*. See section II-1.1 for definitions.

USED FOR INDICATORS:	Op25, Fi28, Fi29
ALSO USED FOR **CI**:	CI69, CI70, CI72

A8 – Billed metered consumption (m^3)

Total amount of billed metered authorised consumption (including exported water) during the assessment period.

INPUT DATA

Referred to a reference period

This input data results from the sum of customer meter readings. As in general readings dates do not refer to the exact audit period, interpolations will be required to have the best possible estimate of the true value.

See section II-1.1 for definitions and Table 79 – Components of water balance specifying the variable codes.

USED FOR INDICATORS:	WR1(A19), Op23(A15), Op24(A15), Op27(A19), Op28(A19), Op39, Fi1(A14), Fi4(A14), Fi5(A14), Fi6(A14), Fi25(A14), Fi28(A14), Fi46(A21), Fi47(A19,G57)
USED FOR VARIABLES:	A10, A14(A10), A15(A14), A19(A15), A20, A21(A20), G57

A9 – Billed unmetered consumption (m^3)

Total amount of billed unmetered authorised consumption (including exported water) during the assessment period.

INPUT DATA

Referred to a reference period

This input data is the best available estimate, based on surveys or any other forms of assessment the *water utility* can make use of.

See section II-1.1 for definitions and Table 79 – Components of water balance specifying the variable codes.

USED FOR INDICATORS:	WR1(A19), Op23(A15), Op24(A15), Op27(A19), Op28(A19), Fi1(A14), Fi4(A14), Fi5(A14), Fi6(A14), Fi25(A14), Fi28(A14), Fi46(A21), Fi47(A19,G57)
USED FOR VARIABLES:	A10, A14(A10), A15(A14), A19(A15), A20, A21(A20), G57

A10 – Billed authorised consumption (m^3)

Total amount of billed authorised consumption (including exported water) during the assessment period.

A8 + A9

Referred to a reference period

Note that billed authorised consumption may include items such as fire fighting and training, flushing of mains and sewers, street cleaning, watering of municipal gardens, public fountains, frost protection, building water, etc., if these are billed.

See section II-1.1 for definitions and Table 79 – Components of water balance specifying the variable codes.

USED FOR INDICATORS:	WR1(A19), Op23(A15), Op24(A15), Op27(A19), Op28(A19), Fi1(A14), Fi4(A14), Fi5(A14), Fi6(A14), Fi25(A14), Fi28(A14), Fi47(A19)
USED FOR VARIABLES:	A14, A15(A14), A19(A15)

A11 – Unbilled metered consumption (m^3)

Total amount of unbilled metered authorised consumption (including exported water) during the assessment period.

INPUT DATA

Referred to a reference period

Note that unbilled metered consumption may include items such as fire fighting and training, flushing of mains and sewers, street cleaning, watering of municipal gardens, public fountains, frost protection, building water, etc, if these are unbilled and metered.

See section II-1.1 for definitions and Table 79 – Components of water balance specifying the variable codes.

USED FOR INDICATORS:	WR1(A19), Op23(A15), Op24(A15), Op27(A19), Op28(A19), Op39, Fi1(A14), Fi4(A14), Fi5(A14), Fi6(A14), Fi25(A14), Fi28(A14), Fi47(A13,A19)
USED FOR VARIABLES:	A13, A14(A13), A15(A14), A19(A15)

A12 – Unbilled unmetered consumption (m^3)

Total amount of unbilled unmetered authorised consumption (including exported water) during the assessment period.

INPUT DATA

Referred to a reference period

This input data is the best available estimate, based on surveys or any other forms of assessment the *water utility* can make use of.

Note that unbilled unmetered consumption may include items such as fire fighting and training, flushing of mains and sewers, street cleaning, watering of municipal gardens, public fountains, frost protection, building water, etc., if these are unbilled and unmetered.

See section II-1.1 for definitions and Table 79 – Components of water balance specifying the variable codes.

USED FOR INDICATORS:	WR1(A19), Op23(A15), Op24(A15), Op27(A19), Op28(A19), Fi1(A14), Fi4(A14), Fi5(A14), Fi6(A14), Fi25(A14), Fi28(A14), Fi47(A13,A19)
USED FOR VARIABLES:	A13, A14(A13), A15(A14), A19(A15)

A13 – Unbilled authorised consumption (m^3)

Total amount of unbilled authorised consumption (including exported water) during the assessment period.

A11 + A12

Referred to a reference period

Note that unbilled metered consumption may include items such as fire fighting and training, flushing of mains and sewers, street cleaning, watering of municipal gardens, public fountains, frost protection, building water, etc., if unbilled. These may be, metered or unmetered, according to local practice.

See section II-1.1 for definitions and Table 79 – Components of water balance specifying the variable codes.

Used for indicators:	WR1(A19), Op23(A15), Op24(A15), Op27(A19), Op28(A19), Fi1(A14), Fi4(A14), Fi5(A14), Fi6(A14), Fi25(A14), Fi28(A14), Fi47
Used for variables:	A14, A15(A14), A19(A15)

A14 – Authorised consumption (m^3)

Total volume of metered and/or non-metered water that, during the assessment period, is taken by registered customers, by the water supplier itself, or by others who are implicitly or explicitly authorised to do so by the water supplier, for residential, commercial, industrial or public purposes. It includes water exported.

A10 + A13

Referred to a reference period

Note that authorised consumption may include items such as fire fighting and training, flushing of mains and sewers, street cleaning, watering of municipal gardens, public fountains, frost protection, building water, etc. These may be billed or unbilled, metered or unmetered, according to local practice.

See section II-1.1 for definitions and Table 79 – Components of water balance specifying the variable codes.

Used for indicators:	WR1(A19), Op23(A15), Op24(A15), Op27(A19), Op28(A19), Fi1, Fi4, Fi5, Fi6, Fi25, Fi28, Fi47(A19)
Used for variables:	A15, A19(A15)
Also used for CI:	CI69, CI72, CI76

A15 – Water losses (m^3)

Difference between system input volume and authorised consumption

A3 – A14

Referred to a reference period

Water losses can be considered as a total volume for the whole system, or for partial systems such as transmission or distribution. In each case the components of the calculation would be adjusted accordingly. Water losses consist of real losses and apparent losses.

See section II-1.1 for definitions and Table 79 – Components of water balance specifying the variable codes.

Used for indicators:	WR1(A19), Op23, Op24, Op27(A19), Op28(A19), Fi47(A19)
Used for variables:	A19

A16 – Unauthorised consumption (m^3)

Total amount of unauthorised water consumption during the assessment period, including water theft.

Input data

Referred to a reference period

The input data is the best available estimate, based on surveys or any other forms of assessment the *water utility* can make use of.

See section II-1.1 for definitions and Table 79 – Components of water balance specifying the variable codes.

Used for indicators:	WR1(A19), Op25(A18), Op26(A18), Op27(A19), Op28(A19), Fi47(A18)
Used for variables:	A18, A19(A18)

A17 – Metering inaccuracies water losses (m^3)

Total amount of water consumed during the assessment period, but unaccounted-for due to metering inaccuracies.

Input data

Referred to a reference period

The input data is the best available estimate, based primarily on data from the calibration of existing meters and/or surveys held for this scope.

See section II-1.1 for definitions and Table 79 – Components of water balance specifying the variable codes.

Used for indicators:	WR1(A19), Op25(A18), Op26(A18), Op27(A19), Op28(A19), Fi47(A18)
Used for variables:	A18, A19(A18)

A18 – Apparent losses (m^3)

Total amount of water unaccounted-for due to unauthorised consumption and metering inaccurancies, during the assessment period.

A16 + A17

Referred to a reference period

See section II-1.1 for definitions and Table 79 – Components of water balance specifying the variable codes.

Apparent losses do not include errors in system input meters. However, it is important to calculate the impact of these errors in commertial terms.

USED FOR INDICATORS: WR1(A19), Op25, Op26, Op27(A19), Op28(A19), Fi47

USED FOR VARIABLES: A19

A19 – Real losses (m^3)

Total amount of physical water losses from the pressurised system during the assessment period, up to the point of customer metering.

A15 – A18

Referred to a reference period

The volume lost during the assessment period through all types of leaks, bursts and overflows depends on frequencies, flow rates, and average duration of individual leaks.

See section II-1.1 for definitions and Table 79 – Components of water balance specifying the variable codes.

USED FOR INDICATORS: WR1, Op27, Op28, Fi47

A20 – Revenue water (m^3)

Total amount of billed authorised consumption (including exported water) during the assessment period.

A8 + A9

Referred to a reference period

The same as A10 – Billed authorised consumption.

See section II-1.1 for definitions and Table 79 – Components of water balance specifying the variable codes.

USED FOR INDICATORS: Fi46(A21)

USED FOR VARIABLES: A21

A21 – Non-revenue water (m^3)

Difference between the system input volume and the billed authorised consumption (including exported water) during the assessment period.

A3 – A20

Referred to a reference period

Non-revenue water includes not only the real and apparent losses but also the unbilled authorised consumption.

It is recommended that if the term unaccounted-for water is used, it should be defined and calculated in the same way as non-revenue water. Accordingly, no specific performance indicators for UFW have been recommended in this manual.

See section II-1.1 for definitions and Table 79 – Components of water balance specifying the variable codes.

USED FOR INDICATORS: Fi46

A22 – Reused supplied water (m^3)

Total volume of wastewater directly reused as water source during the assessment period, after adequate treatment, generally as a complement of conventional water sources.

INPUT DATA

Referred to a reference period

USED FOR INDICATORS: WR4

3.2 Section B – Personnel data

B1 – Total personnel (No.)

Total number of full time equivalent employees of the *water utility* at the reference date.

B2 + B3 + B4 + B5 + B6

Referred to a reference date

In the case of multi-utilities, each employee shall be accounted for based on the proportion of time dedicated to water supply related activities.

Employees may include permanent and temporary personnel.

USED FOR INDICATORS: Pe1, Pe2, Pe3, Pe4, Pe5, Pe6, Pe7, Pe8, Pe9, Pe16, Pe17, Pe18, Pe19, Pe20, Pe21, Pe22, Pe23, Pe24, Pe25

USED FOR VARIABLES: B17

ALSO USED FOR CI: CI6

B2 – General management personnel (No.)

Total number of full time equivalent employees dedicated to directorate, central administration, strategic planning, marketing and comunications, other stakeholder relations, legal affairs, internal audits, environmental management, new business development and general computing support at the reference date.

INPUT DATA

Referred to a reference date

In the case of multi-utilities, each employee shall be accounted for based on the proportion of time dedicated to water supply related activities.

Employees may include permanent and temporary personnel.

See definitions of section II-1.2 – Organisation functions.

USED FOR INDICATORS: Pe1(B1), Pe2(B1), Pe3, Pe4(B1), Pe5(B1), Pe6(B1), Pe7(B1), Pe8(B1), Pe9(B1), Pe16(B1), Pe17(B1), Pe18(B1), Pe19(B1), Pe20(B1), Pe21(B1), Pe22(B1), Pe23(B1), Pe24(B1), Pe25(B1)

USED FOR VARIABLES: B1, B17(B1)

B3 – Human resources management personnel (No.)

Total number of full time equivalent employees of the *water utility* dedicated to personnel administration, education and training, occupational safety, and health services and social activities, at the reference date.

INPUT DATA

Referred to a reference date

In the case of multi-utilities, each employee shall be accounted for based on the proportion of time dedicated to water supply related activities.

Employees may include permanent and temporary personnel.

See definitions of section 1.2 – Organisation functions.

USED FOR INDICATORS: Pe1(B1), Pe2(B1), Pe3(B1), Pe4, Pe5(B1), Pe6(B1), Pe7(B1), Pe8(B1), Pe9(B1), Pe16(B1), Pe17(B1), Pe18(B1), Pe19(B1), Pe20(B1), Pe21(B1), Pe22(B1), Pe23(B1), Pe24(B1), Pe25(B1)

USED FOR VARIABLES: B1

B4 – Financial and commercial personnel (No.)

Total number of full time equivalent employees dedicated to economic and financial planning, economic administration, economic controlling and purchasing and material management activities, at the reference date.

INPUT DATA

Referred to a reference date

In the case of multi-utilities, each employee shall be accounted for based on the proportion of time dedicated to water supply related activities.

Employees may include permanent and temporary personnel.

See definitions of section II-1.2 – Organisation functions.

USED FOR INDICATORS:	Pe1(B1), Pe2(B1), Pe3(B1), Pe4(B1), Pe5, Pe6(B1), Pe7(B1), Pe8(B1), Pe9(B1), Pe16(B1), Pe17(B1), Pe18(B1), Pe19(B1), Pe20(B1), Pe21(B1), Pe22(B1), Pe23(B1), Pe24(B1), Pe25(B1)
USED FOR VARIABLES:	B1, B17(B1)

B5 – Customer service personnel (No.)

Total number of full time equivalent employees dedicated to accounting and control and to customer relations and management activities, at the reference date.

INPUT DATA

Referred to a reference date

In the case of multi-utilities, each employee shall be accounted for based on the proportion of time dedicated to water supply related activities.

Employees may include permanent and temporary personnel.

See definitions of section II-1.2 – Organisation functions.

For the ease of assessment, this variable can be broken down into more detailed sub-variables.

USED FOR INDICATORS:	Pe1(B1), Pe2(B1), Pe3(B1), Pe4(B1), Pe5(B1), Pe6, Pe7(B1), Pe8(B1), Pe9(B1), Pe16(B1), Pe17(B1), Pe18(B1), Pe19(B1), Pe20(B1), Pe21(B1), Pe22(B1), Pe23(B1), Pe24(B1), Pe25(B1)
USED FOR VARIABLES:	B1, B17(B1)

B6 – Technical services personnel (No.)

Total number of full time equivalent employees dedicated to planning, construction, operations and maintenance activities, at the reference date.

B9 + B10 + B11 + B12 + B13 + B14, OR **B7 + B8** OR, IF THESE VARIABLES ARE NOT AVAILABLE, INPUT DATA

Referred to a reference date

In the case of multi-utilities, each employee shall be accounted for based on the proportion of time dedicated to water supply related activities.

Employees may include permanent and temporary personnel.

See definitions of section II-1.2 – Organisation functions.

USED FOR INDICATORS: Pe1(B1), Pe2(B1), Pe3(B1), Pe4(B1), Pe5(B1), Pe6(B1), Pe7, Pe8(B1), Pe9(B1), Pe16(B1), Pe17(B1), Pe18(B1), Pe19(B1), Pe20(B1), Pe21(B1), Pe22(B1), Pe23(B1), Pe24(B1), Pe25(B1)

USED FOR VARIABLES: B1, B17(B1)

B7 – Planning and construction personnel (No.)

Total number of full time equivalent employees of technical services working in planning & construction of water supply systems, at the reference date.

INPUT DATA

Referred to a reference date

In the case of multi-utilities, each employee shall be accounted for based on the proportion of time dedicated to water supply related activities.

Employees may include permanent and temporary personnel.

See definitions of section II-1.2 – Organisation functions.

USED FOR INDICATORS: Pe1(B1), Pe2(B1), Pe3(B1), Pe4(B1), Pe5(B1), Pe6(B1), Pe7(B6), Pe8, Pe9(B1), Pe16(B1), Pe17(B1), Pe18(B1), Pe19(B1), Pe20(B1), Pe21(B1), Pe22(B1), Pe23(B1), Pe24(B1), Pe25(B1)

USED FOR VARIABLES: B1(B6), B6, B17(B1)

B8 – Operations and maintenance personnel (No.)

Total number of full time equivalent employees of technical services working in operations & maintenance of water supply systems, at the reference date.

INPUT DATA

Referred to a reference date

In the case of multi-utilities, each employee shall be accounted for based on the proportion of time dedicated to water supply related activities.

Employees may include permanent and temporary personnel.

See definitions of section II-1.2 – Organisation functions.

USED FOR INDICATORS:	Pe1(B1), Pe2(B1), Pe3(B1), Pe4(B1), Pe5(B1), Pe6(B1), Pe7(B6), Pe8(B1), Pe9, Pe16(B1), Pe17(B1), Pe18(B1), Pe19(B1), Pe20(B1), Pe21(B1), Pe22(B1), Pe23(B1), Pe24(B1), Pe25(B1)
USED FOR VARIABLES:	B1(B6), B6, B17(B1)

B9 – Water resources and catchment management personnel (No.)

Total number of full time equivalent employees working in the water resources and catchment management, at the reference date.

INPUT DATA

Referred to a reference date

Employees may include permanent and temporary personnel.

See definitions of section II-1.2 – Organisation functions.

USED FOR INDICATORS:	Pe1(B1), Pe2(B1), Pe3(B1), Pe4(B1), Pe5(B1), Pe6(B1), Pe7(B6), Pe8(B1), Pe9(B1), Pe10, Pe16(B1), Pe17(B1), Pe18(B1), Pe19(B1), Pe20(B1), Pe21(B1), Pe22(B1), Pe23(B1), Pe24(B1), Pe25(B1)
USED FOR VARIABLES:	B1(B6), B6, B17(B1)

B10 – Abstraction and treatment personnel (No.)

Total number of full time equivalent employees working in the planning, design, construction, operations and maintenance of the water abstraction and treatment, at the reference date.

INPUT DATA

Referred to a reference date

Employees may include permanent and temporary personnel.

See definitions of section II-1.2 – Organisation functions.

For the ease of assessment, this variable can be broken down into more detailed sub-variables.

USED FOR INDICATORS: Pe1(B1), Pe2(B1), Pe3(B1), Pe4(B1), Pe5(B1), Pe6(B1), Pe7(B6), Pe8(B1), Pe9(B1), Pe11, Pe16(B1), Pe17(B1), Pe18(B1), Pe19(B1), Pe20(B1), Pe21(B1), Pe22(B1), Pe23(B1), Pe24(B1), Pe25(B1)

USED FOR VARIABLES: B1(B6), B6, B17(B1)

B11 – Transmission, storage and distribution personnel (No.)

Total number of full time equivalent employees working in planning, design, construction, operations and maintenance of the transmission, storage and distribution system, at the reference date.

INPUT DATA

Referred to a reference date

Employees may include permanent and temporary personnel.

See definitions of section II-1.2 – Organisation functions.

For the ease of assessment, this variable can be broken down into more detailed sub-variables.

USED FOR INDICATORS: Pe1(B1), Pe2(B1), Pe3(B1), Pe4(B1), Pe5(B1), Pe6(B1), Pe7(B6), Pe8(B1), Pe9(B1), Pe12, Pe16(B1), Pe17(B1), Pe18(B1), Pe19(B1), Pe20(B1), Pe21(B1), Pe22(B1), Pe23(B1), Pe24(B1), Pe25(B1)

USED FOR VARIABLES: B1(B6), B6, B17(B1)

B12 – Water quality monitoring personnel (No.)

Total number of full time equivalent employees working in water quality sampling and testing, at the reference date.

INPUT DATA

Referred to a reference date

In the case of multi-utilities, each employee shall be accounted for based on the proportion of time dedicated to water supply related activities.

Employees may include permanent and temporary personnel.

See definitions of section II-1.2 – Organisation functions.

For the ease of assessment, this variable can be broken down into more detailed sub-variables.

USED FOR INDICATORS: Pe1(B1), Pe2(B1), Pe3(B1), Pe4(B1), Pe5(B1), Pe6(B1), Pe7(B6), Pe8(B1), Pe9(B1), Pe13, Pe16(B1), Pe17(B1), Pe18(B1), Pe19(B1), Pe20(B1), Pe21(B1), Pe22(B1), Pe23(B1), Pe24(B1), Pe25(B1)

USED FOR VARIABLES: B1(B6), B6, B17(B1)

B13 – Meter management personnel (No.)

Total number of full time equivalent employees working in the installation, maintenance and replacement of water meters, at the reference date.

INPUT DATA

Referred to a reference date

Customer meter reading personnel must not be included because it is part of commercial and customer service. System flow meter reading personnel (e.g., district metering) shall be included.

Employees may include permanent and temporary personnel.

See definitions of section II-1.2 – Organisation functions.

USED FOR INDICATORS: Pe1(B1), Pe2(B1), Pe3(B1), Pe4(B1), Pe5(B1), Pe6(B1), Pe7(B6), Pe8(B1), Pe9(B1), Pe14, Pe16(B1), Pe17(B1), Pe18(B1), Pe19(B1), Pe20(B1), Pe21(B1), Pe22(B1), Pe23(B1), Pe24(B1), Pe25(B1)

USED FOR VARIABLES: B1(B6), B6, B17(B1)

B14 – Support services personnel (No.)

Total number of full time equivalent employees working in the central stock repository, central workshops and the vehicle fleet, at the reference date.

INPUT DATA

Referred to a reference date

In the case of multi-utilities, each employee shall be accounted for based on the proportion of time dedicated to water supply related activities.

Employees may include permanent and temporary personnel.

See definitions of section II-1.2 – Organisation functions.

For the ease of assessment, this variable can be broken down into more detailed sub-variables.

USED FOR INDICATORS:	Pe1(B1), Pe2(B1), Pe3(B1), Pe4(B1), Pe5(B1), Pe6(B1), Pe7(B6), Pe8(B1), Pe9(B1), Pe15, Pe16(B1), Pe17(B1), Pe18(B1), Pe19(B1), Pe20(B1), Pe21(B1), Pe22(B1), Pe23(B1), Pe24(B1), Pe25(B1)
USED FOR VARIABLES:	B1(B6), B6, B17(B1)

B15 – University degree personnel (No.)

Number of full time equivalent employees of the *water utility* with university degree, at the reference date.

INPUT DATA

Referred to a reference date

In the case of multi-utilities, each employee shall be accounted for based on the proportion of time dedicated to water supply related activities.

Employees may include permanent and temporary personnel.

USED FOR INDICATORS:	Pe16, Pe18(B17)
USED FOR VARIABLES:	B17

B16 – Basic education personnel (No.)

Total number of full-time equivalent employees of the *water utility* without a university degree but with at least a basic education, at the reference date.

INPUT DATA

Referred to a reference date

The definition of basic education to be used shall refer to the country requirements at the indicator assessment date. Definitions may vary from country to country.

In the case of multi-utilities, each employee shall be accounted for based on the proportion of time dedicated to water supply related activities.

Employees may include permanent and temporary personnel.

USED FOR INDICATORS: Pe17, Pe18(B17)

USED FOR VARIABLES: B17

B17 – Other qualification personnel (No.)

Total number of full-time equivalent employees without basic education, at the reference date.

B1 – (B15 + B16)

Referred to a reference date

The definition of basic education to be used shall refer to the country requirements at the indicator assessment date. Definitions may vary from country to country.

In the case of multi-utilities, each employee shall be accounted for based on the proportion of time dedicated to water supply related activities.

Employees may include permanent and temporary personnel.

USED FOR INDICATORS: Pe18

B18 – Total training time (hour)

Total number of training hours during the assessment period.

B19 + B20

Referred to a reference period

In the case of multi-utilities, the training hours provided to each employee shall be accounted for based on the proportion of the time dedicated to water supply activities.

Employees may include permanent and temporary personnel.

USED FOR INDICATORS: Pe19

B19 – Internal training time (hour)
Total number of internal training hours during the assessment period. INPUT DATA Referred to a reference period
In the case of multi-utilities, the training hours provided to each employee shall be accounted for based on the proportion of the time dedicated to water supply activities. Employees may include permanent and temporary personnel. USED FOR INDICATORS: Pe19(B18), Pe20 USED FOR VARIABLES: B18

B20 – External training time (hour)
Total number of external training hours during the assessment period. INPUT DATA Referred to a reference period
In the case of multi-utilities, the training hours provided to each employee shall be accounted for based on the proportion of the time dedicated to water supply activities. Employees may include permanent and temporary personnel. USED FOR INDICATORS: Pe19(B18), Pe21 USED FOR VARIABLES: B18

B21 – Working accidents (No.)
Total number of working accidents requiring medical care occurring with personnel during the assessment period. INPUT DATA Referred to a reference period
In the case of multi-utilities, only the accidents related to water supply activities shall be considered. In the case of employees working in general activities, each working accident shall be accounted for based on the proportion of the time dedicated to water supply activities. Employees may include permanent and temporary personnel. USED FOR INDICATORS: Pe22

B22 – Absenteeism (day)

Total number of days of absenteeism occurring during the assessment period.

B23 + B24

Referred to a reference period

In the case of multi-utilities, the absentism of each employee shall be accounted for based on the proportion of the time dedicated to water supply related activities.

Absenteeism includes all absence to service, excluding vacation leave.

Employees may include permanent and temporary personnel.

Used for indicators: Pe23

B23 – Absenteeism due to accidents or illness at work (days)

Total number of days of absenteeism due to accidents or illness at work occurring during the assessment period.

Input data

Referred to a reference period

In the case of multi-utilities, the absentism of each employee shall be accounted for based on the proportion of the time dedicated to water supply related activities.

Employees may include permanent and temporary personnel.

Used for indicators: Pe23(B22), Pe24

Used for variables: B22

B24 – Absenteeism due to other reasons (day)

Total number of days of absenteeism occurring during the assessment period that were not due to working accidents or illness at work.

Input data

Referred to a reference period

In the case of multi-utilities, the absentism of each employee shall be accounted for based on the proportion of the time dedicated to water supply related activities.

Employees may include permanent and temporary personnel.

Used for indicators: Pe23(B22), Pe25

Used for variables: B22

B25 – Working time (hours)	
Total number of normal working hours of all the *utility* employees during the assessment period (employed hours gross – absence for holidays).	
INPUT DATA	
	Referred to a reference period
In the case of multi-utilities, the working time of each employee shall be accounted for based on the proportion of the time dedicated to water supply related activities. Overtime working hours shall not be included. Employees may include permanent and temporary personnel.	
USED FOR INDICATORS:	Pe26
ALSO USED FOR CI:	CI6

B26 – Overtime work (hours)	
Total number of overtime working hours of the *utility* employees during the assessment period.	
INPUT DATA	
	Referred to a reference period
In the case of multi-utilities, the overtime working hours of each employee shall be accounted for based on the proportion of the time dedicated to water supply related activities. Employees may include permanent and temporary personnel.	
USED FOR INDICATORS:	Pe26

3.3 Section C – Physical assets data

Water storage

C1 – Raw water storage capacity (m^3)	
Total net volume of raw water reservoirs included in the system, at the reference date.	
INPUT DATA	
	Referred to a reference date
In the cases of multi-use reservoirs, the net volume available for the provision of the water supply service shall be used. This variable aims to measure the usable physical capacity of the impounding storage facilities (i.e. raw water reservoirs) regardless of the water availability to fill them in.	
USED FOR INDICATORS:	Ph2
ALSO USED FOR CI:	CI25

C2 – Treated water storage capacity (m^3)	
Total volume of transmission and distribution service reservoirs, at the reference date.	
INPUT DATA	
	Referred to a reference date
The customer storage tanks must not be included. If appropriate, this variable can be split into different components, e.g., equalisation capacity, capacity for fire flow protection and emergency storage.	
USED FOR INDICATORS:	Ph3, Op2
ALSO USED FOR CI:	CI32

Treatment plants

C3 – Daily treatment capacity (m^3/day)	
Maximum daily global capacity of the existing treatment plants, at the reference date.	
INPUT DATA	
	Referred to a reference date
USED FOR INDICATORS:	Ph1

Pumping stations

C4 – Pumps (No.)	
Total number of pumps of the system, at the reference date.	
INPUT DATA	
	Referred to a reference date
When accounting for the total number of pumps of the system, small pumps can be excluded if their influence in terms of the system reliability is negligible.	
USED FOR INDICATORS:	Op30

C5 – Pumping stations (No.)

Total number of pumping stations of the system (customer pumping systems excluded), at the reference date.

INPUT DATA

Referred to a reference date

If a pumping station feeding the system under analysis serves other systems as well, a pumped volume-based proportion of this pumping station shall be used.

USED FOR INDICATORS: Op34

ALSO USED FOR CI: CI33

C6 – Pumping stations capacity (kW)

Total nominal power of the system pumps (customer pumping systems excluded), at the reference date.

INPUT DATA

Referred to a reference date

If a pumping station of the system under analysis serves other systems as well, a pumped volume-based proportion of this pumping station shall be used.

When accounting for the total pumping capacity of the system, the power of small pumps can be excluded if their influence in terms of global confidence grade of the variable is negligible.

USED FOR INDICATORS: Op1, Op21, Op22

ALSO USED FOR CI: CI34

C7 – Maximum operating pumping capacity (kW)

Maximum nominal pumping power that can be used simultaneously in the system.

INPUT DATA

Referred to a reference date

This variable shall exclusively account for the system pumps, including the backup ones, that can be used simultaneously in each pumping station.

If a pumping station of the system under analysis serves other systems as well, a pumped volume-based proportion of this pumping station shall be used.

When accounting for the total pumping capacity of the system, the power of small pumps can be excluded if their influence in terms of global confidence grade of the variable is negligible.

USED FOR INDICATORS: Ph4

VARIABLES

Transmission and distribuition network

C8 – Mains length (km)

Total transmission and distribution mains length (*service connections* not included), at the reference date.

INPUT DATA

Referred to a reference date

Mains that are not yet in use or have been put out of service on a permanent basis shall not be accounted for.

USED FOR INDICATORS: Pe12, Op3, Op4, Op5, Op15, Op16, Op17, Op18, Op24, Op28, Op29, Op31

ALSO USED FOR CI: CI35, CI61, CI76

C9 – Distribution mains length (km)

Total distribution mains length (transmission lines and *service connections* not included), at the reference date.

INPUT DATA

Referred to a reference date

Mains that are not yet in use or have been put out of service on a permanent basis shall not be accounted for.

USED FOR INDICATORS: Ph8, Ph9

Metering and control equipment

C10 – System flow meters (No.)

Total number of flow meters available for permanent or temporary installation in the transmission and distribution system, such as meters for system control, district metering and accounting of imported and exported water (direct customer meters, E6, shall not be included.

INPUT DATA

Referred to a reference date

Meters installed or ready for installation when necessary shall be accounted for. Equipment that is not calibrated or is waiting for *repair* shall not be included.

USED FOR INDICATORS: Pe14, Op7

C11 – District meters (No.)

Total number of water metering points permanently or systematically equipped for district metering, at the reference date.

INPUT DATA

Referred to a reference date

USED FOR INDICATORS: Ph10

C12 – Pressure meters (No.)

Total number of pressure meters available for permanent or temporay pressure monitoring, at the reference date.

INPUT DATA

Referred to a reference date

Meters installed or ready for installation when necessary shall be accounted for. Meters that are not calibrated or are waiting for *repair* shall not be included.

USED FOR INDICATORS: Op9

C13 – Water level meters (No.)

Total number of water level meters available for permanent or temporary water level monitoring, at the reference date.

INPUT DATA

Referred to a reference date

Meters installed or ready for installation when necessary shall be accounted for. Meters that are not calibrated or are waiting for *repair* shall not be included.

USED FOR INDICATORS: Op10

C14 – On-line water quality monitoring instruments (No.)

Total number of instruments available for permanent or temporary on-line water quality monitoring, at the reference date.

INPUT DATA

Referred to a reference date

Instruments installed or ready for installation when necessary shall be accounted for. Instruments that are not calibrated or are waiting for *repair* shall not be included.

USED FOR INDICATORS: Op11

C15 – Control units (No.)

Total number of control units, defined as functionally independent sets of control devices of the system (e.g., control valves or pump switches) used in water wells, surface water catchments, pumping stations, energy recovery stations, treatment plants, storage tanks and raw water, transmission and distribution networks (at the reference date).

INPUT DATA

Referred to a reference date

For example, the set of valves and other control devices used for controlling a group of wells sharing multiple pumping and storage installations shall be considered a single control unit. A similar criterium can be used, for instance, for each unit process in treatment plants and associated pumping and storage facilities (e.g., filtration unit).

USED FOR INDICATORS: Ph14, Ph15

C16 – Automated control units (No.)

Total number of automated control units of the system, defined as functionally independent sets of control devices of the system (e.g., control valves or pump switches) that are locally controlled by a programmable logical controller, used in water wells, surface water catchments, pumping stations, energy recovery stations, treatment plants, storage tanks and raw water, transmission and distribution networks (at the reference date).

INPUT DATA

Referred to a reference date

USED FOR INDICATORS: Ph14

C17 – Remotely controlled units (No.)

Total number of control units of the system remotely controled from a central control system, usually via information cabling or radio connection, used in water wells, surface water catchments, pumping stations, energy recovery stations, treatment plants, storage tanks and raw water, transmission and distribution networks (at the reference date).

INPUT DATA

Referred to a reference date

USED FOR INDICATORS: Ph15

C18 – Emergency power systems (kW)	
Total nominal power of the emergency power systems, at the reference date. INPUT DATA Referred to a reference period	
USED FOR INDICATORS:	Op12

C19 – Signal transmission equipment (No.)	
Total number of signal transmission units, at the reference date. INPUT DATA Referred to a reference date	
USED FOR INDICATORS:	Op13

C20 – Electrical switchgear (No.)	
Total number of electrical switchgear equipments, at the reference date. INPUT DATA Referred to a reference date	
USED FOR INDICATORS:	Op14

C21 – Mains valves (No.)	
Total number of all kinds of valves installed in transmission and distribution system mains (valves installed in the *service connections* shall not be accounted for), at the reference date. INPUT DATA Referred to a reference date	
USED FOR INDICATORS:	Op19

C22 – Isolating valves (No.)	
Total number of isolating valves of all types installed in transmission and distribution system mains (valves installed in the *service connections* shall not be accounted for), at the reference date. INPUT DATA Referred to a reference date	
USED FOR INDICATORS:	Ph8

C23 – Hydrants (No.)

Total number of hydrants of all types installed in the distribution network, at the reference date.

INPUT DATA

Referred to a reference date

All types of hydrants shall be accounted for, including hydrants for flushing.

USED FOR INDICATORS: Ph9, Op6, Op33

Service connections

C24 – Service connections (No.)

Total number of *service connections*, at the reference date.

INPUT DATA

Referred to a reference date

When there is no reliable information about the number of *service connections*, it is suggested that the number of buildings is used as a first estimate. However, this estimate should be corrected taking into account that it is undervaluated due to the fact that some buildings may have more than one connection and connections to consumption points not located in buildings (e.g., irrigation hydrants, public taps) are likely to exist.

USED FOR INDICATORS: Pe1, Ph10, Ph11, Op20, Op23, Op27, Op29, Op32, QS10, QS14, QS26

ALSO USED FOR CI: CI54, CI61, CI72

C25 – Average service connection length (m)

Average length (metres) from the property boundary (delivery point) to the *measurement point*, at the reference date.

INPUT DATA

Referred to a reference date

Frequently water *utilities* do not have detailed accurate information to assess this variable. In these cases, an estimate shall be adopted.

USED FOR INDICATORS: Op29

ALSO USED FOR CI: CI56

3.4 Section D – Operational data

Energy consumption

D1 – Pumping energy consumption (kWh)

Total energy consumption for water pumping (customer pumping systems excluded) during the assessment period.

INPUT DATA

Referred to a reference period

This variable is the sum of the actual energy consumption of every water pumping equipment of the system. It shall be assessed from energy consumption meters.

When accounting for the total energy consumption for pumping, the consumption of small pumps can be excluded if their influence in terms of global confidence grade of the variable is negligible.

USED FOR INDICATORS: Ph5, Ph6, Ph7

D2 – Maximum daily pumping energy consumption (kWh)

Sum, for all installed pumps, of the number of operation hours of the maximum energy consumption day during the assessment period multiplied by the nominal power of the pump, resulting in the maximum daily pumping energy consumption of the system.

INPUT DATA

Referred to a reference period

The consumption of small pumps can be excluded if their influence in terms of global confidence grade of this variable is negligible.

USED FOR INDICATORS: Ph4

D3 – Standardisation factor ($m^3 \times 100m$)

Sum of D3(i) for all the pumps of the system, D3(i) being:

D3(i) = V(i) × h(i)/100, where V is the total volume (m^3) pumped by pump i during the assessment period and h(i) is the pump head (m).

INPUT DATA

Referred to a reference period

For pumps with significant variation of pump head throughout the assessment period, it may be necessary to break down the this period into a limited number of time intervals. For instance, if a pump works 1/3 of the time with a flow of 10 m^3/h and a pump head of 50 m, and 2/3 of the time with a flow of 12 m^3/h and a pump head of 42 m, D3(i) will be:

$$D3(i) = ((10 \times 24 \times 365/3) \times 50 + (12 \times 24 \times 365 \times 2/3) \times 42)/100$$

The contribution of small pumps can be excluded if their influence in terms of global confidence grade of the variable is negligible.

USED FOR INDICATORS: Ph5

D4 – Reactive energy consumption (kVar)

Total reactive energy consumption for pumping in the system during the assessment period.

INPUT DATA

Referred to a reference period

In general, reactive energy consumption is assessed from specific surveys. It shall be reduced to a value as close to zero as possible by the installation of adequate compensation equipment.

The contribution of small pumps can be excluded if their influence in terms of global confidence grade of the variable is negligible.

USED FOR INDICATORS: Ph6

D5 – Energy recovery (Wh)

Total energy recovered by the use of turbines or reverse pumps in the system during the assessment period.

INPUT DATA

Referred to a reference period

Energy recovery may be cost-effective in gravity transmission lines where the potential energy available is excessive for the hydraulic transport needs.

USED FOR INDICATORS: Ph7

Inspection and maintenance

D6 – Pumping inspection (kW)

Sum of the nominal power of all pumps and related ancillaries in the system subjected to *inspection* during the assessment period.

INPUT DATA

Referred to a reference period

Pumps integrated in the treatment process and customer pumping systems are excluded.

If a pump is inspected more than once during the assessment period, it shall be accounted as many times as the number of inspections.

The contribution of small pumps can be excluded if their influence in terms of global confidence grade of the variable is negligible.

USED FOR INDICATORS: Op1

D7 – Storage tank cleaning (m^3)

Total volume of water storage tanks cells cleaned during the assessment period.

INPUT DATA

Referred to a reference period

Cleaning means the implementation of a formal procedure, as a rule written, the results of which are subsequently recorded.

If a cell is cleaned more than once during the assessment period, it shall be accounted as many times as the number of cleanings.

USED FOR INDICATORS: Op2

D8 – Network inspection (km)

Length of transmission and distribution mains where at least valves and other fittings were *inspected* during the assessment period.

INPUT DATA

Referred to a reference period

Inspection means the implementation of a formal procedure, as a rule written, the results of which are subsequently recorded in a form that enables the *utility* to assess the serviceability of the assets and to take corrective action as appropriate.

If a pipe is inspected more than once during the assessment period, it shall be accounted only once. This convention may cause bias to the indicator when a non-annual assessment period is adopted.

USED FOR INDICATORS: Op3

VARIABLES

D9 – Leakage control (km)

Length of mains subject to *active leakage control* during the assessment period.

INPUT DATA

Referred to a reference period

USED FOR INDICATORS: Op4

D10 – Leaks repaired due to active leakage control (No.)

Number of leaks detected and *repaired* due to *active leakage control* during the assessment period.

INPUT DATA

Referred to a reference period

USED FOR INDICATORS: Op5

D11 – Hydrant inspection (No.)

Total number of hydrants *inspected* during the assessment period.

INPUT DATA

Referred to a reference period

If a hydrant is inspected more than once during the assessment period, it shall be accounted as many times as the number of inspections.

USED FOR INDICATORS: Op6

D12 – System flow meter calibrations (No.)

Number of system flowmeter *calibrations* carried out during the assessment period.

INPUT DATA

Referred to a reference period

Only the *calibrations* properly recorded by the *utility* shall be accounted for.

If a system flow meter is calibrated more than once during the assessment period, it shall be accounted as many times as the number of calibrations.

USED FOR INDICATORS: Op7

D13 – Pressure meter calibrations (No.)

Number of pressure meter *calibrations* carried out during the assessment period.

INPUT DATA

Referred to a reference period

Only the *calibrations* properly recorded by the *utility* shall be accounted for.

If a pressure meter is calibrated more than once during the assessment period, it shall be accounted as many times as the number of calibrations.

USED FOR INDICATORS: Op9

D14 – Water level meter calibrations (No.)

Number of water level meter *calibrations* carried out during the assessment period.

INPUT DATA

Referred to a reference period

Only the *calibrations* properly recorded by the *utility* shall be accounted for.

If a water level meter is calibrated more than once during the assessment period, it shall be accounted as many times as the number of calibrations.

USED FOR INDICATORS: Op10

D15 – On-line water quality monitoring equipment calibrations (No.)

Number of *calibrations* carried out during the assessment period on on-line water quality monitoring instruments installed in the system (permanently or temporarily).

INPUT DATA

Referred to a reference period

Only the *calibrations* properly recorded by the *utility* shall be accounted for.

If an on-line water quality monitoring instrument is calibrated more than once during the assessment period, it shall be accounted as many times as the number of calibrations.

USED FOR INDICATORS: Op11

D16 – Emergency power systems inspection (kW)

Sum of the nominal power of the system emergency power equipment *inspected* during the assessment period.

INPUT DATA

Referred to a reference period

Inspection means the implementation of a formal procedure, as a rule written, the results of which are subsequently recorded in a form that enables the *utility* to assess the serviceability of the assets and to take corrective action as appropriate.

If a given equipment is *inspected* more than once during the assessment period, it shall be accounted for as many times as the number of inspections.

USED FOR INDICATORS: Op12

D17 – Signal transmission equipment inspection (No.)

Number of signal transmission units in the system *inspected* during the assessment period.

INPUT DATA

Referred to a reference period

Inspection means the implementation of a formal procedure, as a rule written, the results of which are subsequently recorded in a form that enables the *utility* to assess the serviceability of the assets and to take corrective action as appropriate.

If a given equipment is *inspected* more than once during the assessment period, it shall be accounted for as many times as the number of inspections.

USED FOR INDICATORS: Op13

D18 – Electrical switchgear inspection (No.)

Number of electrical switchgear units in the system *inspected* during the assessment period.

INPUT DATA

Referred to a reference period

Inspection means the implementation of a formal procedure, as a rule written, the results of which are subsequently recorded in a form that enables the *utility* to assess the serviceability of the assets and to take corrective action as appropriate.

If a given equipment is *inspected* more than once during the assessment period, it shall be accounted for as many times as the number of inspections.

USED FOR INDICATORS: Op14

D19 – Permanent vehicles (No.)
Number of permanent vehicles daily available, in average, for field works in operations and maintenance activities, at the reference date. INPUT DATA Referred to a reference date
In the case of multi-utilities, only the portion of vehicles related to water supply activities shall be considered. All vehicles that are available for the use of the *water utility* personnel for field works in operations and maintenance activities shall be accounted for, regardless of their ownership. Vehicles used in the scope of outsourced activities shall be excluded. Heavy equipment shall not be included. USED FOR INDICATORS: Op15

Preventive maintenance

D20 – Mains rehabilitation (km)
Length of transmission and distribution mains rehabilitated during the assessment period. INPUT DATA Referred to a reference period
This variable includes not only D21 and D22 but also the length of mains rehabilitated with other techniques. USED FOR INDICATORS: Op16

D21 – Mains renovation (km)
Length of mains renovated during the assessment period by epoxy resine, cement mortar or other materials. INPUT DATA Referred to a reference period
See section 1.4 – Complementary definitions for the definition of *renovation*. USED FOR INDICATORS: Op17

VARIABLES

D22 – Mains replacement (km)

Mains length replaced during the assessment period.

INPUT DATA

Referred to a reference period

See section 1.4 for the definition of *replacement.*

USED FOR INDICATORS: Op18

D23 – Replaced valves (No.)

Number of mains valves replaced during the assessment period.

INPUT DATA

Referred to a reference period

Valves installed in *service connections* shall not be included.

USED FOR INDICATORS: Op19

D24 – Service connection rehabilitation (No.)

Number of *service connections* replaced or renovated during the assessment period.

INPUT DATA

Referred to a reference period

USED FOR INDICATORS: Op20

D25 – Pumps overhaul (kW)

Total nominal power of pumps subject, during the assessment period, to refurbishment of relevant elements, necessary to bring unit back to original performance.

INPUT DATA

Referred to a reference period

Overhaul may include partial replacement.

The power of small pumps can be excluded if their influence in terms of global confidence grade of the variable is negligible.

USED FOR INDICATORS: Op21

D26 – Pumps replacement (kW)

Total nominal power of pumps replaced during the assessment period.

INPUT DATA

Referred to a reference period

Pump replacement refers exclusively to full replacement of the pump and the engine. The nominal power to be accounted for shall refer to the new equipment.

The power of small pumps can be excluded if their influence in terms of global confidence grade of the variable is negligible.

USED FOR INDICATORS: Op22

Failures

D27 – Pump failures (day)

Sum, for all pumps, of the number of days during the assessment period when the pump is out of order.

INPUT DATA

Referred to a reference period

When accounting for the total number of pump failures, small pumps can be excluded if their influence in terms of the system reliability is negligible.

USED FOR INDICATORS: Op30

D28 – Mains failures (No.)

Number of *mains failures* during the assessment period, including failures of valves and fittings.

INPUT DATA

Referred to a reference period

For ease of assessment, mains failures can be considered equivalent to the number of recorded main *repairs*, assuming that all failures detected are *repaired* and recorded.

If mains failures are to be used for regulating objectives, the use of a complementary indicator, similar to Op31 but excluding failures by third parties is advisable, as they are not a direct fault of the *water utility*. This variable shall exclude *repairs* under active leakage control.

USED FOR INDICATORS: Op31

D29 – Service connection failures (No.)

Number of *service connection* failures during the assessment period.

INPUT DATA

Referred to a reference period

For ease of assessment, *service connection* failures can be considered equivalent to the number of recorded *service connection repairs*, assuming that all failures detected are *repaired* and recorded.

If *service connection* failures are to be used for regulating objectives, the use of a complementary indicator, similar to Op32 but excluding failures by third parties is advisable, as they are not a direct fault of the *water utility*. This variable shall exclude *repairs* under active leakage control.

USED FOR INDICATORS: Op32

D30 – Hydrant failures (No.)

Number of hydrant failures during the assessment period.

INPUT DATA

Referred to a reference period

If hydrant failures are to be used for regulating objectives, the use of a complementary indicator, similar to Op33 but excluding failures by third parties is advisable, as they are not a direct fault of the *water utility*. This variable shall exclude *repairs* under active leakage control.

USED FOR INDICATORS: Op33

D31 – Power failures (hour)

Sum, for all water pumping stations, of the number of hours during the assessment period each pumping station is out of service due to power supply interruptions.

INPUT DATA

Referred to a reference period

This variable assesses the time that pumping stations are out of service due to power failures, regardless of the power source used (normal or standby power generation system). This means that the time when a pumping station is reliant on standby power generation shall not be considered as a power failure. This indicator aims to measure the performance of the water supply *utility* and not of the power supply *utility*.

USED FOR INDICATORS: Op34

VARIABLES

D32 – Water-point failures (No.)

Number of water-points failures during the reference period.

INPUT DATA

Referred to a reference period

This variable is relevant for developing regions. Water point refers to public taps and standpipes. Water-point failures shall be accounted for regardless of their cause (tap, pump, piping, water source, etc.).

USED FOR INDICATORS: Op35

Network pressure and service continuity

D33 – Delivery points with adequate pressure (No.)

Number of *delivery points* that receive and are likely to receive pressure equal to or above the guaranteed or declared target level at the peak demand hour (but not when demand is *abnormal*), at the reference date.

INPUT DATA

Referred to a reference date

This variable can be applied also in the case of bulk supply. In this case the it shall read "number of *delivery points* that are supplied at any time according to the target flow, volume and/or pressure".

USED FOR INDICATORS: QS10, QS11

D34 – Average operating pressure (kPa)

Average operating pressure at the *delivery points* when system is pressurised, at the reference date.

INPUT DATA

Referred to a reference date

An exact measure of the average operating pressure would require the continuous monitoring of the pressure at every *delivery point*. In practice, simplifications are required. In flat regions, an estimate of this variable is easy to get. Also when calibrated extended period hydraulic models are available, nodal pressures weighted with the nodal demand can be adopted. In hilly areas where no better estimates can be obtained, simplified pressure contour maps can be drawn and an estimate for the average operating pressure in each contour band is established. The global value is than assessed as a weighted-average of the pressure at each contour band, using their respective equivalent-population as weighting factor.

USED FOR INDICATORS: Op29

D35 – Water interruptions (person x hour)

Sum, for all the water interruptions, of the population subject to a water interruption multiplied by the duration of the interruption in hours), during the assessment period.

INPUT DATA

Referred to a reference period

In this context, only the unplanned (even if notified) or un-notified water supply interruption to customers with a duration (measured to full restoration of supply) of more than 12 hours, caused by bursts or failures in the water supply system and the subsequent *repair*/renewal measures, shall be accounted for. Includes those planned interruptions that exceed the notified period.

USED FOR INDICATORS: QS13

D36 – Service interruptions (No.)

Total number of water service interruptions, during the assessment period.

INPUT DATA

Referred to a reference period

In this context, only the unplanned (even if notified) or un-notified water supply interruption to customers with a duration (measured to full restoration of supply) of more than 12 hours, caused by bursts or failures in the water supply system and the subsequent *repair*/renewal measures, shall be accounted for. Includes also those planned interruptions that exceed the notified period.

Interruptions inherent to a systematic intermittent supply must not be accounted in this variable.

USED FOR INDICATORS: QS14, QS15

D37 – Water use restrictions (person x hour)

Sum, for all the *restrictions to water service* occurred during the assessment period, of the population affected by each restriction to water service multiplied by the respective duration in hours.

INPUT DATA

Referred to a reference period

USED FOR INDICATORS: QS16

D38 – Days with restrictions to water service (day)
Total number of days with *restrictions to water service* during the assessment period.
INPUT DATA
Referred to a reference period
USED FOR INDICATORS: QS17

Metering

D39 – Residential customer meter reading frequency (No./meter/year)
Frequency of residential customer meter readings, pre-defined by the *water utility* for the assessment period.
INPUT DATA
Referred to a reference period
'Residential' includes all customers that have the same procedures for meter reading as the metered residential consumers (i.e. commercial).
USED FOR INDICATORS: Op36, Op37

D40 – Industrial customer meter reading frequency (No./meter/year)
Frequency of industrial customer readings meter, pre-defined by the *water utility* for the assessment period.
INPUT DATA
Referred to a reference period
'Industrial' includes all customers that have the same procedures for meter reading as the metered industrial consumers (i.e. with meters above a given diameter or above a given average consumption).
USED FOR INDICATORS: Op36

D41 – Bulk customer meter reading frequency (No./meter/year)

Frequency of bulk customer readings meter, pre-defined by the *water utility* for the assessment period.

INPUT DATA

Referred to a reference period

'Bulk' includes all customers that achieve water to the *water utility* to deliver to a third party. They correspond to the raw and treated water exports.

USED FOR INDICATORS: Op36

D42 – Customer meter readings (No.)

Total number of *effective meter readings* carried out during the assessment period for all types of metered customers.

INPUT DATA

Referred to a reference period

USED FOR INDICATORS: Op36

D43 – Residential customer meter readings (No.)

Total number of *effective meter readings* carried out during the assessment period for metered residential customers.

INPUT DATA

Referred to a reference period

This variable refers to residential meters and to any other meters that have the same procedures for meter reading as the metered residential consumers.

USED FOR INDICATORS: Op37

D44 – Operational meters (No.)

Number of direct customer meters that are not out-of-service at the reference time.

INPUT DATA

Referred to a reference date

This variable shall include the direct customer meters that are producing readings used for accounting purposes. Meters that are not working or that produce readings that are not used due to insufficient confidence grades shall not be accounted for.

USED FOR INDICATORS: Op38

D45 – Meter replacement (No.)	
Number of customer meters replaced in the system during the assessment period.	
INPUT DATA	
	Referred to a reference period
USED FOR INDICATORS:	Op8

Water quality monitoring

D46 – Required treated water quality tests carried out (No.)	
Number of *treated water* tests carried out during the assessment period that are required by applicable standards or legislation.	
D47 + D48 + D49 + D50	
	Referred to a reference period
For each parameter, the number of tests carried out to be accounted for this variable cannot exceed the number of tests that are required by applicable standards or legislation.	
USED FOR INDICATORS:	Op40

D47 – Required aesthetic tests carried out (No.)	
Number of aesthetic tests of *treated water* carried out during the assessment period that are required by applicable standards or legislation.	
INPUT DATA	
	Referred to a reference period
For each parameter, the number of tests carried out to be accounted for this variable cannot exceed the number of tests that are required by applicable standards or legislation.	
USED FOR INDICATORS:	Op40(D46), Op41
USED FOR VARIABLES:	D46

VARIABLES

D48 – Required microbiological tests carried out (No.)

Number of microbiological tests of *treated water* carried out during the assessment period that are required by applicable standards or legislation.

INPUT DATA

Referred to a reference period

For each parameter, the number of tests carried out to be accounted for this variable cannot exceed the number of tests that are required by applicable standards or legislation.

USED FOR INDICATORS: Op40(D46), Op42

USED FOR VARIABLES: D46

D49 – Required physical-chemical tests carried out (No.)

Number of physical-chemical tests of *treated water* carried out during the assessment period that are required by applicable standards or legislation.

INPUT DATA

Referred to a reference period

For each parameter, the number of tests carried out to be accounted for this variable cannot exceed the number of tests that are required by applicable standards or legislation.

USED FOR INDICATORS: Op40(D46), Op43

USED FOR VARIABLES: D46

D50 – Required radioactivity tests carried out (No.)

Number of radioactivity tests of *treated water* carried out during the assessment period that are required by applicable standards or legislation.

INPUT DATA

Referred to a reference period

For each parameter, the number of tests carried out to be accounted for this variable cannot exceed the number of tests that are required by applicable standards or legislation.

USED FOR INDICATORS: Op40(D46), Op44

USED FOR VARIABLES: D46

D51 – Treated water quality tests carried out (No.)

Number of *treated water* tests carried out during the assessment period.

D53 + D54 + D55 + D56

Referred to a reference period

For each parameter, all treated water tests carried out shall be accounted for, even though the number of tests that are required by applicable standards or legislation is exceeded.

USED FOR INDICATORS: QS18

D52 – Water quality tests carried out (No.)

Total number of tests carried out by the *water utility* laboratories during the assessment period.

INPUT DATA

Referred to a reference period

This variable includes not only the treated water tests but also raw water and process control tests. Outsourced tests shall not be included.

USED FOR INDICATORS: Pe13

D53 – Aesthetic tests carried out (No.)

Number of aesthetic tests of *treated water* carried out during the assessment period.

INPUT DATA

Referred to a reference period

For each parameter, all treated water tests carried out shall be accounted for, even though the number of tests that are required by applicable standards or legislation is exceeded.

USED FOR INDICATORS: QS18(D51), QS19

USED FOR VARIABLES: D51

D54 – Microbiological tests carried out (No.)

Number of microbiological tests of *treated water* carried out during the assessment period.

INPUT DATA

Referred to a reference period

For each parameter, all treated water tests carried out shall be accounted for, even though the number of tests that are required by applicable standards or legislation is exceeded.

USED FOR INDICATORS: QS18(D51), QS20

USED FOR VARIABLES: D51

D55 – Physical-chemical tests carried out (No.)

Number of physical-chemical tests of *treated water* carried out during the assessment period.

INPUT DATA

Referred to a reference period

For each parameter, all treated water tests carried out shall be accounted for, even though the number of tests that are required by applicable standards or legislation is exceeded.

USED FOR INDICATORS: QS18(D51), QS21

USED FOR VARIABLES: D51

D56 – Radioactivity tests carried out (No.)

Number of radioactivity tests of *treated water* carried out during the assessment period.

INPUT DATA

Referred to a reference period

For each parameter, all *treated water* tests carried out shall be accounted for, even though the number of tests that are required by applicable standards or legislation is exceeded.

USED FOR INDICATORS: QS18(D51), QS22

USED FOR VARIABLES: D51

D57 – Water quality tests required (No.)

Number of *treated water* tests required by applicable standards or legislation during the assessment period.

D58 + D59 + D60 + D61

Referred to a reference period

Standards may be of general use, be defined by the *utility* or result from specific contractual agreements.

USED FOR INDICATORS: Op40

D58 – Aesthetic tests required (No.)

Number of aesthetic tests of *treated water* required by applicable standards or legislation during the assessment period.

INPUT DATA

Referred to a reference period

Standards may be of general use, be defined by the *utility* or result from specific contractual agreements.

USED FOR INDICATORS: Op40(D57), Op41

USED FOR VARIABLES: D57

D59 – Microbiological tests required (No.)

Number of microbiological tests of *treated water* tests required by applicable standards or legislation during the assessment period.

INPUT DATA

Referred to a reference period

Standards may be of general use, be defined by the *utility* or result from specific contractual agreements.

USED FOR INDICATORS: Op40(D57), Op42

USED FOR VARIABLES: D57

D60 – Physical-chemical tests required (No.)

Number of physical-chemical tests of *treated water* required by applicable standards or legislation during the assessment period.

INPUT DATA

Referred to a reference period

Standards may be of general use, be defined by the *utility* or result from specific contractual agreements.

USED FOR INDICATORS: Op40(D57), Op43

USED FOR VARIABLES: D57

D61 – Radioactivity tests required (No.)

Number of radioactivity tests of *treated water* required by applicable standards or legislation during the assessment period.

INPUT DATA

Referred to a reference period

Standards may be of general use, be defined by the *utility* or result from specific contractual agreements.

USED FOR INDICATORS: Op40(D57), Op44

USED FOR VARIABLES: D57

D62 – Compliance of aesthetic tests (No.)

Number of aesthetic tests of *treated water* carried out during the assessment period, which comply with the applicable standards or legislation.

INPUT DATA

Referred to a reference period

Standards may be of general use, be defined by the *utility* or result from specific contractual agreements.

USED FOR INDICATORS: QS18, QS19

D63 – Compliance of microbiological tests (No.)

Number of microbiological tests of *treated water* carried out during the assessment period, which comply with the applicable standards or legislation.

INPUT DATA

Referred to a reference period

Standards may be of general use, be defined by the *utility* or result from specific contractual agreements.

USED FOR INDICATORS: QS18, QS20

D64 – Compliance of physical-chemical tests (No.)

Number of physical-chemical tests of *treated water* carried out during the assessment period, which comply with the applicable standards or legislation.

INPUT DATA

Referred to a reference period

Standards may be of general use, be defined by the *utility* or result from specific contractual agreements.

USED FOR INDICATORS: QS18, QS21

D65 – Compliance of radioactivity tests (No.)

Number of radioactivity tests of *treated water* carried out during the assessment period, which comply with the applicable standards or legislation.

INPUT DATA

Referred to a reference period

Standards may be of general use, be defined by the *utility* or result from specific contractual agreements.

USED FOR INDICATORS: QS18, QS22

3.5 Section E – Demography and customer data

E1 – Households and businesses supplied (No.)

Total number of households and businesses connected to the water supply system at the reference date.

INPUT DATA

Referred to a reference date

USED FOR INDICATORS: QS1

E2 – Buildings supplied (No.)

Total number of buildings connected to the water supply system at the reference date.

INPUT DATA

Referred to a reference date

USED FOR INDICATORS: QS2

E3 – Households and businesses (No.)

Total number of household and businesses of the area of influence of the *water utility* regarding water supply, at the reference date.

INPUT DATA

Referred to a reference date

Whenever the national surveys consider this type of data, the official number should be used. Intermediate estimates published between census questionnaire are considered valid.

USED FOR INDICATORS: QS1

E4 – Buildings (No.)

Total number of buildings of the area of influence of the *water utility* regarding water supply, at the reference date.

INPUT DATA

Referred to a reference date

Whenever the national surveys consider this type of data, the official number should be used. Intermediate estimates published between census questionnaire are considered valid.

USED FOR INDICATORS: QS2

E5 – Resident population (person)

Total population who lives on a permanent basis in the area served by the *water utility*, at the reference date.

INPUT DATA

Referred to a reference date

Whenever the national surveys consider this type of data, the official number should be used. Intermediate estimates published between census questionnaires are considered valid.

USED FOR INDICATORS: QS3, QS4, QS5

ALSO USED FOR **CI:** CI77

E6 – Direct customer meters (No.)

Total number of direct customer water meters at the reference date.

INPUT DATA

Referred to a reference date

Direct customers include residential, industrial, or any other types of existing customers (e.g., commercial, public, institutional), excluding bulk supply customers.

USED FOR INDICATORS: Pe14, Ph11, Ph12, Op8, Op38

E7 – Residential customer meters (No.)

Total number of residential customer water meters at the reference date.

INPUT DATA

Referred to a reference date

'Residential' includes all customers that have the same procedures for meter reading as the metered residential consumers (i.e. commercial).

USED FOR INDICATORS: Ph13, Op36, Op37

E8 – Industrial customer meters (No.)

Total number of industrial customer water meters at the reference date.

INPUT DATA

Referred to a reference date

'Industrial' includes all customers that have the same procedures for meter reading as the metered industrial consumers (i.e. with meters above a given diameter or above a given average consumption).

USED FOR INDICATORS: Op36

E9 – Bulk customer meters (No.)

Total number of bulk customer water meters at the reference date.

INPUT DATA

Referred to a reference date

'Bulk' includes all customers that achieve water to the *water utility* to deliver to a third party. Bulk customer meters correspond to the delivery points of raw and treated water exports.

USED FOR INDICATORS: Ph12, Op36, QS11, QS15

E10 – Registered customers (customer)

Total number of registered water supply customers at the reference date.

INPUT DATA

Referred to a reference date

Registered customers include residential, industrial, bulk and other registered customers.

'Residential' includes all customers that have the same type of contract with the *water utility* as the residential consumers (i.e. businesses).

'Industrial' includes all customers that have the same type of contract with the *water utility* as the metered industrial consumers (i.e. with meters above a given diameter or above a given average consumption).

'Bulk' includes all customers that achieve water to the *water utility* to deliver to a third party. Bulk customers correspond to the delivery points of raw and treated water exports.

'Other' includes all the registered customers not accounted for as 'residential', 'industrial' or 'bulk', such as commercial, public and institutional.

USED FOR INDICATORS: Ph12, QS27, QS32, QS33

ALSO USED FOR **CI**: CI11

E11 – Residential registered customers (customer)

Total number of residential registered water supply customers at the reference date.

INPUT DATA

Referred to a reference date

'Residential' includes all customers that have the same type of contract with the *water utility* as the residential consumers (i.e. businesses).

USED FOR INDICATORS: Ph13

ALSO USED FOR **CI**: CI12

3.6 Section F – Quality of service data

Service

F1 – Population supplied (person)

Resident population served by the *water utility* at the reference date.

F2 + F3

Referred to a reference date

USED FOR INDICATORS: QS3, QS13, QS16

ALSO USED FOR CI: CI70

F2 – Population supplied with service pipes (person)

***Resident population* served by the *water utility* through *service connections* at the reference date.**

INPUT DATA

Referred to a reference date

This variable is relevant only for developing regions, where a significant part of the supply is provided by public taps and standpipes, and therefore it is relevant to split the supplied population into population supplied (F1) with service pipes (F2) and population served by public taps or standpipes (F3).

USED FOR INDICATORS: QS3(F1), QS4, QS13(F1), QS16(F1)

USED FOR VARIABLES: F1

ALSO USED FOR CI: CI9

F3 – Population served by public taps or standpipes (person)

***Resident population* served by the *water utility* by public taps or standpipes at the reference date.**

INPUT DATA

Referred to a reference date

This variable is relevant only for developing regions where a significant part of the supply is provided by public taps and standpipes.

USED FOR INDICATORS: QS3(F1), QS5, QS8, QS9, QS13(F1), QS16(F1)

USED FOR VARIABLES: F1

F4 – Distance from water points to households (m)

Sum, for all water points, of the distance between the water point and the far-most household served by it, at the reference date.

INPUT DATA

Referred to a reference date

This variable is relevant only for developing regions where a significant part of the supply is provided by public taps and standpipes.

Water points refer to consumption points where one or more public taps or standpipes are located.

USED FOR INDICATORS: QS7

F5 – Public taps and standpipes consumption (m^3)

Sum, for all public taps and standpipes, of the respective water consumption during the assessment period.

INPUT DATA

Referred to a reference period

Whenever feasible, this variable shall be assessed based on flow metering data referred to the consumption points. Sample surveys can be used. If this data is not available, volumes of water supplied to the area can be used, but an estimate of the water losses in the distribution network shall be discounted.

This variable is relevant only for developing regions where a significant part of the supply is provided by public taps and standpipes.

USED FOR INDICATORS: QS8

F6 – Water-points (No.)

Total number of water-points at the reference date.

INPUT DATA

Referred to a reference date

Water-points refer to consumption points where one or more public taps or standpipes are located. All existing waterpoints shall be accounted for, regardless of being operational or not.

This variable is relevant only for developing regions where a significant part of the supply is provided by public taps and standpipes.

USED FOR INDICATORS: Op35, QS6, QS7

F7 – Operational water-points (No.)

Total number of water-points that are not out of service at the reference date.

INPUT DATA

Referred to a reference date

Water points refer to consumption points where one or more public taps or standpipes are located. A water point shall be considered out-of-service if it cannot supply any water to the consumers due to asset failures (e.g., broken taps, out-of-order pumps).

This variable is relevant only for developing regions where a significant part of the supply is provided by public taps and standpipes.

USED FOR INDICATORS: QS6

F8 – Public taps and standpipes (No.)

Total number of public taps and standpipes, at the reference date.

INPUT DATA

Referred to a reference date

Whenever a public tap or standpipe is out of service at the reference data it shall not be accounted for.

This variable is relevant only for developing regions where a significant part of the supply is provided by public taps and standpipes.

USED FOR INDICATORS: QS9

F9 – New connections establishment time (day)

Sum, for all new connections established during the reference period, of the total time spent from the customer request until the availability of the water service.

INPUT DATA

Referred to a reference period

This variable refers to new contracts, when *service connections* already exist.

USED FOR INDICATORS: QS23

F10 – New connections established (No.)

Total number of new connections established during the assessment period.

INPUT DATA

Referred to a reference period

This variable refers to new water service contracts, when *service connections* already exist.

USED FOR INDICATORS: QS23

VARIABLES

F11 – Customer meter installation time (day)

Sum, for all new customer meters installed during the assessment period, of the total time spent from the customer request until the availability of the service.

INPUT DATA

Referred to a reference period

This variable refers exclusively to existing connections that have been newly fitted with a meter, that generally require some construction works.

USED FOR INDICATORS: QS24

F12 – New customer meters installed (No.)

Total number of new water customer meters installed during the assessment period.

INPUT DATA

Referred to a reference period

This variable refers exclusively to existing connections that have been newly fitted with a meter, that therefore generally require some construction works.

USED FOR INDICATORS: QS24

F13 – Connections repair time (day)

Sum, for all *service connections* repaired during the assessment period, of the total time spent from the time failures are reported until the service is resumed.

INPUT DATA

Referred to a reference period

This variable includes all reported service connection failures and not only the failures reported by customers.

USED FOR INDICATORS: QS25

F14 – Connections repaired (No.)

Total number of water connections repaired during the assessment period.

INPUT DATA

Referred to a reference period

This variable includes all connections *repaired* and not only the repairs related to failures reported by customers.

USED FOR INDICATORS: QS25

Customer complaints

F15 – Service complaints (No.)

Number of direct, telephone, and written complaints of quality of service during the assessment period.

F16 + F17 + F18 + F19

Referred to a reference period

In the case of multi-utilities, only the service complaints related to water supply activities shall be considered.

USED FOR INDICATORS: QS26, QS27, QS28, QS29, QS30, QS31

F16 – Pressure complaints (No.)

Number of pressure complaints during the assessment period.

INPUT DATA

Referred to a reference period

USED FOR INDICATORS: QS26(F15), QS28, QS29(F15), QS30(F15), QS31(F15)

USED FOR VARIABLES: F15

F17 – Continuity complaints (No.)

Number of continuity complaints regarding the water supply service during the assessment period.

INPUT DATA

Referred to a reference period

This variable refers to medium- or long-term supply constraints. Complaints on supply interruptions or water use restrictions due to insufficient water quantity or quality at source, insufficient system capacity or on-going works shall be accounted for by this variable. Any other complaints on interruptions shall be included in F19.

USED FOR INDICATORS: QS26(F15), QS28(F15), QS29, QS30(F15), QS31(F15)

USED FOR VARIABLES: F15

F18 – Water quality complaints (No.)

Number of water quality complaints during the assessment period.

INPUT DATA

Referred to a reference period

USED FOR INDICATORS: QS26(F15), QS27(F15), QS28(F15), QS29, QS30(F15), QS31(F15)

USED FOR VARIABLES: F15

F19 – Complaints on interruptions (No.)

Number of complaints due to water supply interruptions during the assessment period.

INPUT DATA

Referred to a reference period

This variable refers to short-term supply interruptions, due to accidental system failures or *repair* works. Supply interruptions due to intermittent supply, insufficient quantity of adequate water at sources, or asset insufficient capacity shall be accounted for in F17 and not in F19.

USED FOR INDICATORS: QS26(F15), QS27(F15), QS28(F15), QS29(F15), QS30(F15), QS31

USED FOR VARIABLES: F15

F20 – Billing complaints and queries (No.)

Number of direct, telephone, and written billing complaints and queries regarding the water supply service, during the assessment period.

INPUT DATA

Referred to a reference period

A billing query is any enquiry regarding a bill (e.g., requests for clarification of the bill contents).

USED FOR INDICATORS: QS32

F21 – Other complaints and queries (No.)

Number of other complaints and queries regarding the water supply service, during the assessment period.

INPUT DATA

Referred to a reference period

USED FOR INDICATORS: QS33

F22 – Written responses (No.)

Number of written responses to written queries and complaints within the target time regarding the water supply service, during the assessment period.

INPUT DATA

Referred to a reference period

USED FOR INDICATORS: QS34

F23 – Written complaints (No.)
Number of written complaints regarding the water supply service, during the assessment period.
INPUT DATA
Referred to a reference period
A written complaint is any letter, note, annotated document, telex, fax, e-mail or any other form of written communication, however mildly worded, that draws attention to any service provided by or action taken by the company or its representatives which falls short of the expectation of the correspondent. All complaints, including those about general levels of charging or other policy issues, and complaints that are not justified must be included.
USED FOR INDICATORS: QS34

3.7 Section G – Economic and financial data

G1 – Total revenues (EUR)
Total operating revenues minus capitalised costs of self-constructed assets, regarding the water supply service, during the assessment period.
G2 – G35
Referred to a reference period
Capitalised costs of self-constructed assets shall be considered as an economic correction of the operating costs. Consequently it should be entered as a negative quantity in order to obtain the total revenues.
See diagram of section II-1.3 – Financial definitions, Table 68 and particularly Table 69, in which some details are given. Exchange rate of local currencies shall be referred to the end of the assessment period.
USED FOR INDICATORS: Fi2, Fi3, Fi30, Fi31

G2 – Operating revenues (EUR)
Total operating revenues during the assessment period, including sales revenues (G3), work in progress, capitalised costs of self-constructed assets (G35) and other operating revenues, regarding the water supply service.
INPUT DATA
Referred to a reference period
See diagram of section II-1.3 – Financial definitions, Table 68 and particularly Table 69, in which some details are given. Exchange rate of local currencies shall be referred to the end of the year.
USED FOR INDICATORS: Fi1, Fi2(G1), Fi3(G1), Fi30(G1), Fi31(G1), Fi38
USED FOR VARIABLES: G1

G3 – Sales revenues (EUR)

Operating revenues from sales during the assessment period, regarding the water supply service.

INPUT DATA

Referred to a reference period

See diagram of section II-1.3 – Financial definitions, Table 68 and particularly Table 69, in which some details are given. Exchange rate of local currencies shall be referred to the end of the year.

USED FOR INDICATORS: Fi2, Fi3, Fi32, Fi45

G4 – Total costs (EUR)

Total costs during the assessment period, including capital and running costs, regarding the water supply service.

G5 + G6

Referred to a reference period

See diagram of section II-1.3 – Financial definitions, Table 68 and Table 69, in which some details are given. Exchange rate of local currencies shall be referred to the end of the year.

USED FOR INDICATORS: Fi4, Fi30

G5 – Running costs (EUR)

Total operations and maintenance net costs and internal manpower net costs (i.e. not including the capitalised cost of self constructed assets) during the assessment period, regarding the water supply service.

G7 + G8

Referred to a reference period

This variable must be equivalent to the sum of G8, G9, G10, G11, G12, G13, G14, G15 and G16, which break down the running costs per type of cost.

See diagram of section II-1.3 – Financial definitions, Table 68 and Table 69. Exchange rate of local currencies shall be referred to the end of the year.

USED FOR INDICATORS: Fi4(G4), Fi5, Fi7, Fi8, Fi9, Fi10, Fi11, Fi12, Fi13, Fi14, Fi15, Fi16, Fi17, Fi18, Fi19, Fi20, Fi21, Fi22, Fi30(G4), Fi31, Fi47

USED FOR VARIABLES: G4

G6 – Capital costs (EUR)

Total net interest and depreciation (based on book values) during the assessment period, regarding the water supply service.

G28 + G31

Referred to a reference period

The net value of interest has to be considered. The interest income shall be accounted for as a reduction in capital costs and not as a revenue.

See diagram of section II-1.3 – Financial definitions, Table 68 and Table 69. Exchange rate of local currencies shall be referred to the end of the year.

USED FOR INDICATORS:	Fi4(G4), Fi6, Fi23, Fi24, Fi30(G4)
USED FOR VARIABLES:	G4

G7 – Operational costs (EUR)

Operational costs during the assessment period regarding the water supply service, including the net value (obtained by negative apportionment of related capitalised costs of self-constructed assets) of: imported (raw and treated) water, energy, external services, leasing and rentals, chemicals, other consumables and materials for maintenance and repair, taxes, levies & fees and exceptional earnings & losses, and other operating costs. Manpower shall not be included.

G9 + G10 + G11 + G12 + G13 + G14 + G15 + G16

Referred to a reference period

See diagram of section II-1.3 – Financial definitions, Table 68 and Table 69. Exchange rate of local currencies shall be referred to the end of the year.

USED FOR INDICATORS:	Fi4(G4), Fi5(G5), Fi7(G5), Fi8(G5), Fi9(G5), Fi10(G5), Fi11(G5), Fi12(G5), Fi13(G5), Fi14(G5), Fi15(G5), Fi16(G5), Fi17(G5), Fi18(G5), Fi19(G5), Fi20(G5), Fi21(G5), Fi22(G5), Fi30(G4), Fi31(G5), Fi47(G5)
USED FOR VARIABLES:	G4(G5), G5

G8 – Internal manpower costs (EUR)

Internal manpower costs during the assessment period regarding the water supply service, referred to the net value (obtained by negative apportionment of related capitalised costs of self-constructed assets) of the personnel which salary is paid directly by the *water utility*.

INPUT DATA

Referred to a reference period

See diagram of section II-1.3 – Financial definitions, Table 68 and Table 69. Manpower costs include salaries and other costs that derive directly from them, such as complementary payments or social security complements. Payment of extra-time work shall be included. Exchange rate of local currencies shall be referred to the end of the year.

USED FOR INDICATORS: Fi4(G4), Fi5(G5), Fi7, Fi8(G5), Fi9(G5), Fi10(G5), Fi11(G5), Fi12(G5), Fi13(G5), Fi14(G5), Fi15(G5), Fi16(G5), Fi17(G5), Fi18(G5), Fi19(G5), Fi20(G5), Fi21(G5), Fi22(G5), Fi30(G4), Fi31(G5), Fi47(G5)

USED FOR VARIABLES: G4(G5), G5

G9 – External services costs (EUR)

The net value (obtained by negative apportionment of related capitalised costs of self-constructed assets) of total costs of external services (i.e. outsourcing), external manpower costs included, regarding the water supply service, during the assessment period.

INPUT DATA

Referred to a reference period

See diagram of II-1.3 – Financial definitions, Table 68 and Table 69. Exchange rate of local currencies shall be referred to the end of the year.

USED FOR INDICATORS: Fi4(G4), Fi5(G5), Fi7(G5), Fi8, Fi9(G5), Fi10(G5), Fi11(G5), Fi12(G5), Fi13(G5), Fi14(G5), Fi15(G5), Fi16(G5), Fi17(G5), Fi18(G5), Fi19(G5), Fi20(G5), Fi21(G5), Fi22(G5), Fi30(G4), Fi31(G5), Fi47(G5)

USED FOR VARIABLES: G4(G5), G5(G7), G7

G10 – Imported (raw and treated) water costs (EUR)	
Total cost of the water imported (raw and treated) during the assessment period.	
INPUT DATA	Referred to a reference period
See diagram of section II-1.3 – Financial definitions, Table 68 and particularly Table 69, in which some details are given. Exchange rate of local currencies shall be referred to the end of the year.	
USED FOR INDICATORS:	Fi4(G4), Fi5(G5), Fi7(G5), Fi8(G5), Fi9, Fi10(G5), Fi11(G5), Fi12(G5), Fi13(G5), Fi14(G5), Fi15(G5), Fi16(G5), Fi17(G5), Fi18(G5), Fi19(G5), Fi20(G5), Fi21(G5), Fi22(G5), Fi30(G4), Fi31(G5), Fi47(G5)
USED FOR VARIABLES:	G4(G5), G5(G7), G7

G11 – Electrical energy costs (EUR)	
Total cost of electrical energy (including energy for pumping and for other water supply *utility* activities such as workshops, office buildings energy consumption, laboratories, etc.), within the period of assessment, regarding the water supply service.	
INPUT DATA	Referred to a reference period
See diagram of section II-1.3 – Financial definitions, Table 68 and particularly Table 69, in which some details are given. This variable includes not only the component proportional to the energy consumption but all the other components of energy costs, such as power tariffs and taxes. Electrical energy consumption for other activities different of the water pumping are also included. Exchange rate of local currencies shall be referred to the end of the year.	
USED FOR INDICATORS:	Fi4(G4), Fi5(G5), Fi7(G5), Fi8(G5), Fi9(G5), Fi10, Fi11(G5), Fi12(G5), Fi13(G5), Fi14(G5), Fi15(G5), Fi16(G5), Fi17(G5), Fi18(G5), Fi19(G5), Fi20(G5), Fi21(G5), Fi22(G5), Fi30(G4), Fi31(G5), Fi47(G5)
USED FOR VARIABLES:	G4(G5), G5(G7), G7

G12 – Purchased merchandises (EUR)

The net value (obtained by negative apportionment of related capitalised costs of self-constructed assets) of all materials, chemicals and other consumables not included in external services costs, regarding the water supply service, during the assessment period.

INPUT DATA

Referred to a reference period

This variable refers to purchases of consumables and other materials for technical and non-technical functions. See diagram of section II-1.3 – Financial definitions, Table 68 and particularly Table 69, in which some details are given. Exchange rate of local currencies shall be referred to the end of the year.

USED FOR INDICATORS:	Fi4(G4), Fi5(G5), Fi7(G5), Fi8(G5), Fi9(G5), Fi10(G5), Fi11, Fi12(G5), Fi13(G5), Fi14(G5), Fi15(G5), Fi16(G5), Fi17(G5), Fi18(G5), Fi19(G5), Fi20(G5), Fi21(G5), Fi22(G5), Fi30(G4), Fi31(G5), Fi47(G5)
USED FOR VARIABLES:	G4(G5), G5(G7), G7

G13 – Leasing and rentals (EUR)

The net value (obtained by negative apportionment of related capitalised costs of self-constructed assets) of total cost of leasing and rentals, regarding the water supply service, during the assessment period.

INPUT DATA

Referred to a reference period

See diagram of section II-1.3 – Financial definitions, Table 68 and particularly Table 69, in which some details are given. Exchange rate of local currencies shall be referred to the end of the year.

USED FOR INDICATORS:	Fi4(G4), Fi5(G5), Fi7(G5), Fi8(G5), Fi9(G5), Fi10(G5), Fi11, Fi12(G5), Fi13(G5), Fi14(G5), Fi15(G5), Fi16(G5), Fi17(G5), Fi18(G5), Fi19(G5), Fi20(G5), Fi21(G5), Fi22(G5), Fi30(G4), Fi31(G5), Fi47(G5)
USED FOR VARIABLES:	G4(G5), G5(G7), G7

G14 – Taxes, levies and fees (EUR)

All levies and licences strictly connected with the water supply system's operation to be paid to a governmental or municipal authority (excluding direct taxes on EBT), during the assessment period.

INPUT DATA

Referred to a reference period

See diagram of section II-1.3 – Financial definitions, Table 68 and particularly Table 69, in which some details are given. Exchange rate of local currencies shall be referred to the end of the year.

USED FOR INDICATORS: Fi4(G4), Fi5(G5), Fi7(G5), Fi8(G5), Fi9(G5), Fi10(G5), Fi11, Fi12(G5), Fi13(G5), Fi14(G5), Fi15(G5), Fi16(G5), Fi17(G5), Fi18(G5), Fi19(G5), Fi20(G5), Fi21(G5), Fi22(G5), Fi30(G4), Fi31(G5), Fi47(G5)

USED FOR VARIABLES: G4(G5), G5(G7), G7

G15 – Exceptional earnings and losses (EUR)

Exceptional income or expenditures related to sales/writing-off of fixed assets, regarding the water supply service, during the assessment period.

INPUT DATA

Referred to a reference period

See diagram of section II-1.3 – Financial definitions, Table 68 and particularly Table 69, in which some details are given. Exchange rate of local currencies shall be referred to the end of the year.

USED FOR INDICATORS: Fi4(G4), Fi5(G5), Fi7(G5), Fi8(G5), Fi9(G5), Fi10(G5), Fi11, Fi12(G5), Fi13(G5), Fi14(G5), Fi15(G5), Fi16(G5), Fi17(G5), Fi18(G5), Fi19(G5), Fi20(G5), Fi21(G5), Fi22(G5), Fi30(G4), Fi31(G5), Fi47(G5)

USED FOR VARIABLES: G4(G5), G5(G7), G7

G16 – Other operating costs (EUR)

The net value (obtained by negative apportionment of related capitalised costs of self-constructed assets) of other operating costs, not included in chemicals, other consumables and materials for maintenance and *repair*, leasing and rentals, taxes, levies and fees and exceptional earnings & losses, regarding the water supply service, during the assessment period.

INPUT DATA

Referred to a reference period

See diagram of section II-1.3 – Financial definitions, Table 68 and particularly Table 69, in which some details are given. Exchange rate of local currencies shall be referred to the end of the year.

USED FOR INDICATORS: Fi4(G4), Fi5(G5), Fi7(G5), Fi8(G5), Fi9(G5), Fi10(G5), Fi11, Fi12(G5), Fi13(G5), Fi14(G5), Fi15(G5), Fi16(G5), Fi17(G5), Fi18(G5), Fi19(G5), Fi20(G5), Fi21(G5), Fi22(G5), Fi30(G4), Fi31(G5), Fi47(G5)

USED FOR VARIABLES: G4(G5), G5(G7), G7

G17 – General management running costs (EUR)

Part of the running cost related to the net value (obtained by negative apportionment of related capitalised costs of self-constructed assets) of costs of directorate, central administration, strategic planning, marketing and comunications, other stakeholder relations, legal affairs, internal audits, environmental management, new business development and general computing support activities, regarding the water supply service, during the assessment period.

INPUT DATA

Referred to a reference period

See definitions of section II-1.2 – Organisation functions, Table 62 and Table 63. Exchange rate of local currencies shall be referred to the end of the year.

USED FOR INDICATORS: Fi12

G18 – Human resources management running costs (EUR)
Part of the running cost related to the net value (obtained by negative apportionment of related capitalised costs of self-constructed assets) of costs of personnel administration, education and training, occupational safety, health services activities and social activities, regarding the water supply service, during the assessment period.
INPUT DATA
Referred to a reference period
See definitions of section II-1.2 – Organisation functions, Table 62 and Table 63. Exchange rate of local currencies shall be referred to the end of the year.
USED FOR INDICATORS: Fi13

G19 – Financial and commercial running costs (EUR)
Part of the running cost related to the net value (obtained by negative apportionment of related capitalised costs of self-constructed assets) of costs of economic and financial planning, economic administration, economic controlling and purchasing and material management activities, regarding the water supply service, during the assessment period.
INPUT DATA
Referred to a reference period
See definitions of section II-1.2 – Organisation functions, Table 64. Exchange rate of local currencies shall be referred to the end of the year.
USED FOR INDICATORS: Fi14

G20 – Customer service running costs (EUR)
Part of the running cost related to the net value (obtained by negative apportionment of related capitalised costs of self-constructed assets) of costs of meter reading, accounting and control, customer relations and management activities, regarding the water supply service, during the assessment period.
INPUT DATA
Referred to a reference period
See definitions of section II-1.2 – Organisation functions, Table 65. Exchange rate of local currencies shall be referred to the end of the year.
USED FOR INDICATORS: Fi15

G21 – Planning, design, construction, operations & maintenance running costs (EUR)

Part of the running cost related to the net value (obtained by negative apportionment of related capitalised costs of self-constructed assets) of the water supply technical planning, design, construction, operations and maintenance (including asset *repair*), during the assessment period.

INPUT DATA

Referred to a reference period

Asset self-construction, renovation and replacement shall not be included. See definitions of section II-1.2 – Organisation functions. Exchange rate of local currencies shall be referred to the end of the year.

USED FOR INDICATORS: Fi16

G22 – Water resources and catchment management running costs (EUR)

Part of the running cost related to the net value (obtained by negative apportionment of related capitalised costs of self-constructed assets) of the water resources and catchment management, during the assessment period.

INPUT DATA

Referred to a reference period

Exchange rate of local currencies shall be referred to the end of the year.

USED FOR INDICATORS: Fi17

G23 – Abstraction and treatment running costs (EUR)

Part of the running cost related to the net value (obtained by negative apportionment of related capitalised costs of self-constructed assets) of the abstraction and treatment, during the assessment period.

INPUT DATA

Referred to a reference period

Exchange rate of local currencies shall be referred to the end of the year.

USED FOR INDICATORS: Fi18

G24 – Transmission, storage and distribution running costs (EUR)

Part of the running cost related to the net value (obtained by negative apportionment of related capitalised costs of self-constructed assets) of the transmission, storage and distribution, during the assessment period.

INPUT DATA

Referred to a reference period

Exchange rate of local currencies shall be referred to the end of the year.

USED FOR INDICATORS: Fi19

G25 – Water quality sampling and testing running costs (EUR)

Part of the running cost related to the net value (obtained by negative apportionment of related capitalised costs of self-constructed assets) of the water quality sampling and testing, during the assessment period.

INPUT DATA

Referred to a reference period

Exchange rate of local currencies shall be referred to the end of the year.

USED FOR INDICATORS: Fi20

G26 – Meter management running costs (EUR)

Part of the running cost related to the net value (obtained by negative apportionment of related capitalised costs of self-constructed assets) of the water meter management, during the assessment period.

INPUT DATA

Referred to a reference period

Exchange rate of local currencies shall be referred to the end of the year.

USED FOR INDICATORS: Fi21

G27 – Support services running costs (EUR)

Part of the running cost related to the net value (obtained by negative apportionment of related capitalised costs of self-constructed assets) of the support services, regarding the water supply service, during the assessment period.

INPUT DATA

Referred to a reference period

Exchange rate of local currencies shall be referred to the end of the year.

USED FOR INDICATORS: Fi22

G28 – Depreciation costs (EUR)

Depreciation costs (on book values), regarding the water supply service, during the assessment period.

INPUT DATA

Referred to a reference period

See diagram of section II-1.3 – Financial definitions, Table 68 and particularly Table 69, in which some details are given. Exchange rate of local currencies shall be referred to the end of the year.

USED FOR INDICATORS: Fi4(G6), Fi6(G6), Fi23, Fi24(G6), Fi30(G4), Fi33, Fi36

USED FOR VARIABLES: G4(G6), G6

G29 – Interest expenses costs (EUR)

Interest expenses costs, regarding the water supply service, during the assessment period.

INPUT DATA

Referred to a reference period

See diagram of section II-1.3 – Financial definitions, Table 68 and particularly Table 69, in which some details are given. Exchange rate of local currencies shall be referred to the end of the year.

USED FOR INDICATORS: Fi4(G6), Fi6(G6), Fi23(G6), Fi24, Fi30(G4)

USED FOR VARIABLES: G4(G6), G6(G31), G31

G30 – Interest income (EUR)

Total interest income, regarding the water supply service, during the assessment period.

INPUT DATA

Referred to a reference period

Interest income shall be considered as an economic correction of interest expenses, and not as revenue. Consequently it should be entered as negative quantity in order to obtain the net interest costs.

See diagram of section II-1.3 – Financial definitions, Table 68 and particularly Table 69, in which some details are given. Exchange rate of local currencies shall be referred to the end of the year.

USED FOR INDICATORS: Fi4(G6), Fi6(G6), Fi23(G6), Fi24, Fi30(G4)

USED FOR VARIABLES: G4(G6), G6(G31), G31

G31 – Net interest (EUR)

Interest expenses costs minus interest income, regarding the water supply service, during the assessment period.

G29 – G30

Referred to a reference period

Most times interest expenses are higher than interest income, what means that the net interest is a cost.

See diagram of section II-1.3 – Financial definitions, Table 68 and particularly Table 69, in which some details are given. Exchange rate of local currencies shall be referred to the end of the year.

USED FOR INDICATORS: Fi4(G6), Fi6(G6), Fi23(G6), Fi24(G6), Fi30(G4)

USED FOR VARIABLES: G4(G6), G6

G32 – Investment in water service assets (EUR)

Total cost of the investments in water service (expenditures for plant and equipment), including capitalised cost of self-constructed water service assets (the part of G35 related to water service assets), regarding the water supply service, during the assessment period.

G33 + G34

Referred to a reference period

See diagram of section II-1.3 – Financial definitions.

Water service assets include investment for supporting buildings, vehicles, etc.

Exchange rate of local currencies shall be referred to the end of the year.

USED FOR INDICATORS: Fi25, Fi26, Fi27, Fi34

G33 – Investments for new assets and reinforcement of existing assets (EUR)

Total cost of the investments in water service assets that constitute a new development for the service (new assets and *reinforcement* of existing assets), including capitalised cost of self-construction (part of G35 related to water service assets newly constructed or reinforced), regarding the water supply service, during the assessment period.

INPUT DATA

Referred to a reference period

Exchange rate of local currencies shall be referred to the end of the year.

USED FOR INDICATORS: Fi25(G32), Fi26, Fi27(G32), Fi34(G32)

USED FOR VARIABLES: G32

G34 – Investments for asset replacement and renovation (EUR)

Total cost of the investments related to the *replacement* and *renovation* of existing water supply assets ("like for like", i.e., maintaining the approximately the same functionality of existing infrastructure), including capitalised cost of self-construction (part of G35 related to the replacement or renovation of water service assets), during the assessment period.

INPUT DATA

Referred to a reference period

Exchange rate of local currencies shall be referred to the end of the year.

USED FOR INDICATORS: Fi25(G32), Fi26(G32), Fi27, Fi34(G32)

USED FOR VARIABLES: G32

G35 – Capitalised cost of self-constructed assets (EUR)

Total capitalised cost of self-constructed assets during the assessment period, regarding the water supply service.

INPUT DATA

Referred to a reference period

This variable includes capitalised cost of self-constructed assets regarding the investments both in water service and in intangible assets that constitute a new development for the service and the investments for the replacement of existing assets.

On these terms it represents:

- An increase of the investments externally constructed and therefore included in G2 and other connected data;
- an economic correction of the gross value of Operation & Maintenance and Internal Manpower costs allocated in the Profit and Loss Statement (see section II-1.3 – Financial definitions, Table 68) and therefore enter as a negative quantity to each type of these costs in G7, G8 and other connected data;
- an economic correction of Operating Revenues allocated in the Profit and Loss Statement (see section II-1.3 – Financial definitions, Table 68) and therefore enter as a negative quantity to obtain the Revenues in G1.

See also Table 69, in which some details are given.

Exchange rate of local currencies shall be referred to the end of the year.

USED FOR INDICATORS: Fi1, Fi2(G1), Fi3(G1), Fi30(G1), Fi31(G1)

USED FOR VARIABLES: G1

G36 – Water sales revenue for direct consumption (EUR)

Water sales revenue for direct water consumption during the assessment period.

INPUT DATA

Referred to a reference period

Direct customers include residential, industrial, or any other types of existing customers (e.g., commercial, public, institutional), excluding bulk supply customers.

Exchange rate of local currencies shall be referred to the end of the year.

USED FOR INDICATORS: Fi28, Fi37(G43), Fi47(G57)

USED FOR VARIABLES: G43, G57

G37 – Water sales revenue for exported water (EUR)

Water sales revenue for exported water.

INPUT DATA

Referred to a reference period

See section II-1.1 – Water balance definitions.

Exchange rate of local currencies shall be referred to the end of the year.

USED FOR INDICATORS: Fi29, Fi37(G43)

USED FOR VARIABLES: G43

G38 – Accounts receivable (EUR)

Accounts receivable from drinking water, at the reference date.

INPUT DATA

Referred to a reference date

To assess the annual PI's, the year-end accounts receivable shall be used.

Exchange rate of local currencies shall be referred to the end of the year.

USED FOR INDICATORS: Fi32

VARIABLES

G39 – Investments subject to depreciation (EUR)

Investments for assets subject to depreciation according to the accounting principles generally accepted, regarding the water supply service, during the assessment period.

INPUT DATA

Referred to a reference period

See diagram of section II-1.3 – Financial definitions, Table 70.

Exchange rate of local currencies shall be referred to the end of the year.

USED FOR INDICATORS: Fi33

G40 – Investments financed by the cash flow (EUR)

Investments financed by the cash flow (assessed as the sum of net income, depreciation and working capital variation), regarding the water supply service, during the assessment period.

INPUT DATA

Referred to a reference period

See diagram of section II-1.3 – Financial definitions, Table 71.

Exchange rate of local currencies shall be referred to the end of the year.

USED FOR INDICATORS: Fi34

G41 – Depreciated historical value of water service assets (EUR)

Sum, for all the water service assets, of the depreciation value applied, since the date of purchase, regarding the water supply service.

INPUT DATA

Referred to a reference date

This variable refers to annual values. See diagram of section II-1.3 – Financial definitions, Table 70 and Table 71.

Exchange rate of local currencies shall be referred to the end of the year.

USED FOR INDICATORS: Fi35, Fi42

G42 – Historical value of water service assets (EUR)

Refers to the gross book value of total expenditures at the reference date in infrastructure and non-infrastructure water service assets, regarding the water supply service.

INPUT DATA

Referred to a reference date

This variable refers to annual values. See diagram of section II-1.3 – Financial definitions, Table 70 and Table 71 and Table 72.

Exchange rate of local currencies shall be referred to the end of the year.

USED FOR INDICATORS: Fi35, Fi36, Fi42

G43 – Annual debt from customers (EUR)

Annual debt from customers, regarding the water supply service.

G44 – (G36 + G37)

Referred to a reference period

This variable refers to annual values. Exchange rate of local currencies shall be referred to the end of the year.

USED FOR INDICATORS: Fi37

G44 – Annual amount billed per year (EUR)

Annual amount billed during the year, regarding the water supply service.

INPUT DATA

Referred to a reference period

This variable refers to annual values. Exchange rate of local currencies shall be referred to the end of the year.

USED FOR INDICATORS: Fi37

USED FOR VARIABLES: G43

G45 – Cash-flow (EUR)

Total available is the sum of net income, depreciation and the net value of decrease or increase in working capital, regarding the water supply service, during the assessment period.

INPUT DATA

Referred to a reference period

See diagram of section II-1.3 – Financial definitions, Table 71.

Exchange rate of local currencies shall be referred to the end of the year.

USED FOR INDICATORS: Fi39

G46 – Financial debt service (EUR)

The financial debt service contains the expenditures for interest (G29), the cost of loans, and the principal (=capital) repayment debt instruments, regarding the water supply service, during the assessment period.

INPUT DATA

Referred to a reference date

Exchange rate of local currencies shall be referred to the end of the year.

USED FOR INDICATORS: Fi39

G47 – Total debt (EUR)

Sum of long term liabilities (bonds and long term financial debts) and current liabilities, at the end of the fiscal year, regarding the water supply service.

G52 + G53

Referred to a reference date

Exchange rate of local currencies shall be referred to the end of the year.

USED FOR INDICATORS: Fi40, Fi43(G48)

USED FOR VARIABLES: G48

G48 – Shareholders' equity (EUR)	
Surplus of the asset over the liabilities, at the end of the fiscal year, regarding the water supply service. G50 – G47 Referred to a reference date	
Shareholders' equity includes subscribed share capital, capital reserves, other reserves and net income for the year. See diagram of section II-1.3 – Financial definitions, Table 72. Exchange rate of local currencies shall be referred to the end of the year.	
USED FOR INDICATORS:	Fi40, Fi43

G49 – Current assets (EUR)	
Current assets include cash at bank and in hand, accounts receivable from drinking water, other accounts receivable, inventories and prepaid expenses, at the reference date, regarding the water supply service. INPUT DATA Referred to a reference date	
When this variable refers to annual values, it should be assessed at the end of the fiscal years. See diagram of section II-1.3 – Financial definitions, Table 72 and II-1.3 – Financial definitions, Table 72. Exchange rate of local currencies shall be referred to the end of the year.	
USED FOR INDICATORS:	Fi41

G50 – Total assets (EUR)	
Sum of intangible assets (including goodwill and net value of licences and rights), water service assets (including net value of *water utility* plants and net value of other assets), financial assets (including net value of financial investments) and current assets (G49), regarding the water supply service, at the end of the fiscal year. INPUT DATA Referred to a reference date	
See diagram of section II-1.3 – Financial definitions, Table 72. In the literature, the term 'total assets' is equivalent to 'invested capital'. Exchange rate of local currencies shall be referred to the end of the year.	
USED FOR INDICATORS:	Fi40(G48), Fi43(G48), Fi44, Fi45
USED FOR VARIABLES:	G48

VARIABLES

G51 – Inventories (EUR)

Accounted value of materials to be used in the production process that are in stock in the *utility* at the reference date, regarding the water supply service.

INPUT DATA

Referred to a reference date

When this variable refers to annual values, it should be assessed for the fiscal years.

This variable is a subset of G49. See diagram of section II-1.3 – Financial definitions, Table 72.

Exchange rate of local currencies shall be referred to the end of the year.

USED FOR INDICATORS: Fi38

G52 – Long term liabilities (EUR)

Sum of bonds and long term financial debts at the reference date, regarding the water supply service.

INPUT DATA

Referred to a reference date

Exchange rate of local currencies shall be referred to the end of the fiscal year.

USED FOR INDICATORS: Fi40(G47,G48), Fi43(G48)

USED FOR VARIABLES: G47, G48(G47)

G53 – Current liabilities (EUR)

Current liabilities include accounts payable, current portion of long term debt and miscellaneous current liabilities, at the reference date, regarding the water supply service.

INPUT DATA

Referred to a reference date

When this variable refers to annual values, it should be assessed for the fiscal years. See diagram of section II-1.3 – Financial definitions, Table 72.

Exchange rate of local currencies shall be referred to the end of the year.

USED FOR INDICATORS: Fi40(G47,G48), Fi41, Fi43(G48)

USED FOR VARIABLES: G47, G48(G47)

G54 – Operating income (EUR)

Earning before interest and taxes (EBIT), regarding the water supply service, during the year.

INPUT DATA

Referred to a reference period

This variable refers to annual values. See diagram of section II-1.3 – Financial definitions, Table 72.

Exchange rate of local currencies shall be referred to the end of the year.

USED FOR INDICATORS: Fi42, Fi44

G55 – Operating income related taxes (EUR)

Subset of the total taxes specifically related to the operating income (EBIT), regarding the water supply service, during the year.

INPUT DATA

Referred to a reference period

This variable refers to annual values. See diagram of section II-1.3 – Financial definitions, Table 72.

Exchange rate of local currencies shall be referred to the end of the assessment period.

USED FOR INDICATORS: Fi44

G56 – Net income (EUR)

Net income after interest payment and taxes, regarding the water supply service, at the end of the fiscal year.

INPUT DATA

Referred to a reference period

This variable refers to annual values. Net income represents the final profit after taxes on EBT have been paid. See diagram of section II-1.3 – Financial definitions, Table 68.

Exchange rate of local currencies shall be referred to the end of the year.

USED FOR INDICATORS: Fi43

G57 – Average water charges for direct consumption (EUR/m^3)

Ratio between the water sales revenue for direct consumption and billed water.

G36/(A8 + A9)

Referred to a reference date

This variable is used to assess the cost of non-revenue. The valuation of non-revenue water components is assessed by the sum of separate valuations for unbilled authorised consumption, apparent losses and real losses. In the case of the unbilled consumption and of the apparent losses, any reduction in their volumes will lead to an increase of billed water. Therefore the valuation shall be assessed based on the average water charge for direct consumption.

Exchange rate of local currencies shall be referred to the end of the year.

Used for indicators: Fi47

G58 – Attributed unit cost for real losses (EUR/m^3)

Highest of the variable component of imported water charge or *long run marginal cost (LRMC)* for own sources.

Input data

Referred to a reference date

This variable is used to assess the cost of non-revenue. The valuation of non-revenue water components is assessed by the sum of separate valuations for unbilled authorised consumption, apparent losses and real losses. In the case of the real losses, any reduction in their volume will correspond to a reduction in the produced or imported water and therefore the valuations shall be based on costs, using the highest of: (i) the variable component of imported water charge; (ii) *long run marginal cost (LRMC)* for own sources.

Exchange rate of local currencies shall be referred to the end of the year.

Used for indicators: Fi47

3.8 Section H – Time data

H1 – Assessment period (day)

Period of time adopted for the assessment of the data and of the PI.

INPUT DATA

Referred to a reference period

The IWA PI system aims to be used annually and therefore it is highly recommended that the year is used as the reference assessment period. However, since the *utilities* may need to track the evolution of their performance within the year, the PI system is prepared to accommodate other assessment periods for most indicators.

In this case, and in order to ensure unit coherence and allow for PI comparison, all the PI expressed in terms of time are formulated in such a way that the values calculated for other assessment periods are converted into annual values.

Attention is drawn to the fact that the behaviour of most variables is not uniform during the year, due to random or seasonal effects, or to activity planning. All comparisons based on PI assessed from non-annual data must take this fact into consideration, in order to avoid any bias.

USED FOR INDICATORS: WR2, WR3, Pe2, Pe10, Pe11, Pe13, Pe19, Pe20, Pe21, Pe22, Pe23, Pe24, Pe25, Ph2, Ph3, Op1, Op2, Op3, Op4, Op5, Op6, Op7, Op8, Op9, Op10, Op11, Op12, Op13, Op14, Op16, Op17, Op18, Op19, Op20, Op21, Op22, Op23, Op24, Op30, Op31, Op32, Op33, Op34, Op35, Op36, Op37, QS8, QS12, QS13, QS14, QS15, QS16, QS17, QS26, QS27, QS32, QS33, Fi32

ALSO USED FOR CI: CI70

H2 – Time system is pressurised (hour)

Amount of time of the year the system is pressurised.

INPUT DATA

Referred to a reference period

Interruptions due to unplanned system failures, or to on-going *repair* or rehabilitation works shall not be accounted for by this variable. In most intermittent supply systems, supply interruptions are not simultaneous all over the network. When there are subsystems supplied in different periods, the indicator has to be assessed individually for each subsystem and the result is a weighted average using the number of *service connections* of each subsystem as weighting factor.

USED FOR INDICATORS: Op27, Op28, Op29(Op27), QS12

3.9 Variables alphabetical lists

Water volume data, page II-276

Annual imported water allowance, A2
Annual yield capacity of own resources, A1
Apparent losses, A18
Authorised consumption, A14
Billed authorised consumption, A10
Billed metered consumption, A8
Billed unmetered consumption, A9
Exported raw water, A5
Exported treated water, A7
Maximum water treated daily, A4
Metering inaccuracies water losses, A17
Non-revenue water, A21
Real losses, A19
Reused supplied water, A22
Revenue water, A20
System input volume, A3
Unauthorised consumption, A16
Unbilled authorised consumption, A13
Unbilled metered consumption, A11
Unbilled unmetered consumption, A12
Water losses, A15
Water produced, A6

Personnel data, page II-286

Absenteeism due to accidents or illness at work, B23
Absenteeism due to other reasons, B24
Absenteeism, B22
Abstraction and treatment personnel, B10
Basic education personnel, B16
Customer service personnel, B5
External training time, B20

4. CONTEXT INFORMATION

4.1 Quick reference CI listing

CONTEXT INFORMATION

Impounding reservoir storage (page II-373)	
CI24 –	Number of impounding reservoirs (No.)
CI25 –	Total capacity of impounding reservoirs (m^3)
Treatment plants (page II-373)	
CI26 –	Treatment plants (No.)
CI27 –	Water delivered without treatment (m^3/day)
CI28 –	Water delivered with disinfection only (m^3/day)
CI29 –	Water delivered with conventional treatment (m^3/day)
CI30 –	Water delivered with advanced treatment (m^3/day)
Transmission and distribution storage tank/service reservoirs (page II-374)	
CI31 –	Number of transmission and distribution storage tanks (No.)
CI32 –	Total capacity of transmission and distribution storage tanks (m^3)
Pumping stations (page II-374)	
CI33 –	Number of pumping stations (No.)
CI34 –	Total capacity of pumping stations (kW)
Transmission and distribution network (page II-375)	
CI35 –	Mains length (km)
Mains materials	
	CI36 – Grey cast iron mains (%)
	CI37 – Ductile iron mains (%)
	CI38 – Steel mains (%)
	CI39 – Ferrous mains without adequate corrosion protection, as proportion of total of above material categories (%)
	CI40 – Asbestos cement mains (%)
	CI41 – Polyethylene mains (%)
	CI42 – Polyvinyl chlorine mains (%)
	CI43 – Concrete mains (%)
	CI44 – Other materials mains (%)
Mains diameters	
CI45 –	mains with Dia <= 100/110 mm (%)
CI46 –	mains with 100/110 < Dia < 300/315 mm (%)
CI47 –	mains with Dia >= 300/315 mm (%)

Mains age	
	CI48 – Mains installed after 2001 (%)
	CI49 – Mains installed 1976 – 2000 (%)
	CI50 – Mains installed 1951 – 1975 (%)
	CI51 – Mains installed 1926 – 1950 (%)
	CI52 – Mains installed before 1925 (%)
CI53 –	Average mains age (years)
Service connections (page II-378)	
CI55 –	Average length of service connections to delivery point (m)
CI56 –	Average length of service connections to measurement point (m)
Service connection materials	
	CI57 – Plastic service connections (%)
	CI58 – Steel service connections (%)
	CI59 – Lead service connections (%)
	CI60 – Other materials service connections (%)
CI61 –	Service connection density (No./km)
Private pumping systems and tanks (page II-379)	
CI62 –	Private pumping systems (%)
CI63 –	Private tanks density (%)
Consumption and peak factors (page II-379)	
Average system input (page II-379)	
CI64 –	Daily average input (m^3/day)
Consumption per type of customer (page II-379)	
	CI65 – Residential consumption (%)
	CI66 – Commercial consumption (%)
	CI67 – Public or institutional consumption (%)
	CI68 – Industrial consumption (%)
	CI69 – Bulk water consumption (%)
CI70 –	Total per capita consumption (l per capita/day)
	CI71 – Residential per capita consumption (l per capita/day)
CI72 –	Consumption per service connection (m^3/connection/year)

Peak factors of supplied and exported water (page II-381)	
CI73 –	Monthly peak factor of supplied and exported water (-)
CI74 –	Daily peak factor of supplied and exported water (-)
CI75 –	Hourly peak factor of supplied and exported water (-)
CI76 –	Network delivery rate (m^3/km/year)
Demography and economics (page II-381)	
CI77 –	Population (No.)
CI78 –	Population density (persons/km^2)
CI79 –	Household occupancy (persons/house)
Population growth rate	
CI80 –	Current population growth rate (% per year)
CI81 –	Forecasted population growth rate (% per year)
CI82 –	Gross National Product per capita (EUR per capita/year)
CI83 –	Inflation rate (% per year)
EUR exchange rate	
CI84 –	Annual average EUR exchange rate (local currency/EUR)
CI85 –	EUR exchange rate at the end of the assessment period (local currency/EUR)
Environment (page II-383)	
Annual rainfall (page II-383)	
CI86 –	Annual average rainfall (mm/year)
CI87 –	Annual maximum rainfall (mm/year)
CI88 –	Annual minimum rainfall (mm/year)
Air temperature (page II-383)	
CI89 –	Daily average air temperature (°C)
CI90 –	Daily maximum air temperature (°C)
CI91 –	Daily minimum air temperature (°C)
Topography (page II-384)	
CI92 –	Source average elevation (m)

CI93 –	Maximum delivery elevation (m)
CI94 –	Minimum delivery elevation (m)
Raw water quality – Source types (page II-384)	
CI95 –	Upland surface water sources (%)
CI96 –	Lowland surface water sources (%)
CI97 –	Natural springs and wetlands sources (%)
CI98 –	Well water sources (%)
CI99 –	Borehole water sources (%)
CI100 –	Saline and brackish water sources (%)

4.2 *Utility* profile

CI1 – Geographical scope
Scope of activity of the organisation as a whole.
VALUES ▪ nation ▪ state ▪ region ▪ local One single choice is to be chosen. Specify designation of the area supplied.

CI2 – Type of activity
Scope of activity of the organisation as a whole, beyond the water supply
VALUES ▪ water supply and no other activity ▪ water supply and gas ▪ water supply and wastewater ▪ water supply and district heating ▪ water supply and electricity ▪ water supply and other (specify): ____________ Multiple choices are valid.

CI3 – Type of assets ownership
Ownership of the water supply infrastructure.
VALUES ▪ public ▪ private ▪ mixed One single choice is valid.

CI4 – Type of operations
Type of operational management of the *utility*.
VALUES ▪ public ▪ private ▪ mixed One single choice is valid.

CI5 – Number of water supply systems (No.)
Number of independent water supply systems managed by the *utility*

CI6 – Annual working time per employee (hours)
Average number of working hours per year for a full time employee (employed hours gross – absence for holidays) Variable ratio: B25/B1

CI7 – Water tariffs
Water selling unit costs, including fixed and variable components, specified per type of customer. The average charges should include both fixed and variables components, and to exclude taxes.
VALUES ▪ average water charge (EUR/m^3) for a residential consumption of 200 m^3/year ▪ average water charge (EUR/m^3) for a commercial or industrial consumption exceeding 10 000 m^3/year Water tariffs shall be considered as context information only when they are neither established by the *water utility* nor cost related.

4.3 Service information

CI8 – Type of system
Classification of the system depeding on the type of service it is aimed for.
VALUES ▪ bulk water supply ▪ direct distribution ▪ bulk supply and direct distribution One single choice can be chosen.

CI9 – Population served (No.)
Size of *resident population* directly supplied by the *utility* (i.e., excluding those supplied through exports) within the supply area Also used as variable F1

CI10 – Peak population served (No.)
Maximum number of resident and non-resident persons directly supplied by the *water utility* within the supply area

CI11 – Number of registered customers (No.)
Total number of registered customers Also used as variable E10

CI12 – Residential registered customers (No.)
Number of registered residential customers Also used as variable E11

CI13 – Non-residential registered customers (No.)
Number of registered non-residential customers

CI14 – Supply area (km²)
Area that can or is intended to be served by the distribution network

CI15 – Type of supply
VALUES ▪ full-time supply ▪ intermittent supply One single choice can be chosen.

CI16 – Minimum service pressure (kPa)
Minimum target service pressure at any *delivery point* of the network, required by applicable legislation or standards.

CI17 – Maximum service pressure (kPa)
Maximum target service pressure at any *delivery point* of the network, required by applicable legislation or standards.

4.4 System assets

Water resources

CI18 – Annual abstraction capacity (m^3/year)
Maximum annual allowance of water abstraction for water supply, based on the availability of raw water resources under normal climatic conditions (i.e., the value used in design and abstraction licence if any)

CI19 – Daily abstraction capacity (m^3/day)
Maximum daily allowance of water abstraction for water supply, ditto

CI20 – Reliable annual yield of sources (m^3/year)
Estimated annual reliable yield of water resources under adverse (drought) conditions (i.e., the value used in supply/demand balance evaluation)

CI21 – Reliable daily yield of sources (m^3/day)
Estimated daily reliable yield ditto

CI22 – Intake/raw water transmission mains capacity (m^3/day)
Maximum daily hydraulic water capacity, with the existing assets (may be greater or less than the capacity of water resources)

CI23 – Protection area (km^2)
Total area with land use constraints specifically determined for water quality protection of drinking water sources

Impounding reservoir storage

CI24 – Number of raw water reservoirs (No.)
Number of raw water reservoirs included in the system.

CI25 – Total capacity of raw water reservoirs (m^3)
Net volume of raw water reservoirs included in the system Also used as variable C1
In the cases of multi-use reservoirs, the volume available for the provision of the water supply service shall be used.

Treatment plants

CI26 – Treatment plants (No.)
Number of treatment plants

Delivered water level of treatment

CI27 – Water delivered without treatment (m^3/day)
Water delivered to users without any treatment

CI28 – Water delivered with disinfection only (m^3/day)
Water delivered to users with disinfection only

CI29 – Water delivered with conventional treatment (m^3/day)
Water delivered to users from *conventional treatment* plants

CI30 – Water delivered with advanced treatment (m^3/day)
Water delivered to users from *advanced treatment* plants

Transmission and distribution storage tanks/Service reservoirs

CI31 – Number of transmission and distribution storage tanks (No.)
Number of transmission and distribution storage tanks (customer storage excluded)

CI32 – Total capacity of transmission and distribution storage tanks (m^3)
Volume of transmission and distribution storage tanks (customer storage excluded) Also used as variable C2

Pumping stations

CI33 – Number of pumping stations (No.)
Number of pumping stations of the transmission and distribution system (customer pumping systems excluded) Also used as variable C5

CI34 – Total capacity of pumping stations (kW)
Total nominal power of the transmission and distribution system pumps (customer pumping systems excluded) Also used as variable C6

Transmission and distribution network

CI35 – Mains length (km)
Transmission and distribution mains length (*service connections* excluded) Also used as variable C8

Mains materials

CI36 – Grey cast iron mains (%)
Length of or grey (including spun) cast iron mains/total mains length × 100

CI37 – Ductile iron mains (%)
Length of ductile iron mains/total mains length × 100

CI38 – Steel mains (%)
Length of steel mains/total mains length × 100

CONTEXT INFORMATION

CI39 – Ferrous mains without adequate corrosion protection, as proportion of total of above material categories (%)

Length of iron and steel mains that are subject to internal and/or external corrosion due to inadequate protection/total length of iron and steel mains × 100

CI40 – Asbestos cement mains (%)

Length of asbestos mains/total mains length × 100

CI41 – Polyethylene mains (%)

Length of polyethylene mains/total mains length × 100

CI42 – Polyvinyl chlorine mains (%)

Length of polyvinyl chlorine mains/total mains length × 100

CI43 – Concrete mains (%)

Length of concrete mains/total mains length × 100

CI44 – Other materials mains (%)

Length of other materials mains/total mains length × 100

Mains diameters

CI45 – Mains with Dia <= 100/110 mm (%)

Length of mains with internal/external diameter <= 100/110 mm/total mains length × 100

CONTEXT INFORMATION

CI46 – Mains with 100/110 < Dia < 300/315 mm (%)

Length of mains with 100/110 mm < internal/external diameter < 300/315 mm/total mains length × 100

CI47 – Mains with Dia >= 300/315 mm (%)

Length of mains with internal/external diameter >= 300/315 mm/total mains length × 100

Mains age

CI48 – Mains installed after 2001 (%)

Length of mains laid after 2001 inclusive/total mains length × 100

CI49 – Mains installed 1976–2000 (%)

Length of mains laid between 1976 and 2000 inclusive/total mains length × 100

CI50 – Mains installed 1951–1975 (%)

Length of mains laid between 1951 and 1975 inclusive/total mains length × 100

CI51 – Mains installed 1926 – 1950 (%)

Length of mains laid between 1926 and 1950 inclusive/total mains length × 100

CI52 – Mains installed before 1925 (%)

Length of mains laid before 1925 inclusive/total mains length × 100

CI52 – Mains installed before 1925 (%)
Length of mains laid before 1925 inclusive/total mains length × 100

CI53 – Average mains age (years)
Average mains age for the global supply system based on the age of each mains and its length

Service connections

CI54 – Total number of service connections (No.)
Number of *service connections* Also used as variable C24

CI55 – Average length of service connections to delivery point (m)
Average length of *service connections* from records or sample data up to the *delivery point*

CI56 – Average length of service connections to measurement point (m)
Average length of *service connections* from records or sample data up to the *measurement point* Also used as variable C25

Service connection materials

CI57 – Plastic service connections (%)
Total number of plastic *service connections*/total number of service connections × 100

CI58 – Steel service connections (%)
Number of steel *service connections*/total number of service connections × 100

CI59 – Lead service connections (%)
Number of lead *service connections*/total number of service connections × 100

CI60 – Other materials service connections (%)
Number of other material *service connections*/total number of service connections × 100

CI61 – Service connection density (No./km)
Number of *service connections*/total mains length Variable ratio: C24/C8

Private pumping systems and tanks

CI62 – Private pumping systems (%)
Number of buildings with private pumping systems (e.g., booster pumps)/total number of buildings × 100

CI63 – Private tanks density (%)
Number of buildings with private tanks/total number of buildings × 100

4.5 Consumption and peak factors

Average system input

CI64 – Daily average input (m^3/day)
Input of the transmission system during the assessment period/assessment period

Consumption per type of customer

CI65 – Residential consumption (%)
Residential consumption during the assessment period/authorised consumption (including exported water) × 100

CI66 – Commercial consumption (%)
Commercial consumption during the assessment period/authorised consumption (including exported water) × 100

CI67 – Public or institutional consumption (%)
Public or institutional consumption during the assessment period/authorised consumption (including exported water) × 100

CI68 – Industrial consumption (%)
Industrial consumption during the assessment period/authorised consumption (including exported water) × 100

CI69 – Bulk water consumption (%)
Exported water during the assessment period/authorised consumption (including exported water) × 100 Variable ratio: (A5 + A7)/A14

CI70 – Total per capita consumption (l per capita/day)
(System input volume – exported water)/population served/assessment period Variable ratio: (A3 – A5 – A7)/F1/H1 × 1000

CI71 – residential per capita consumption (l per capita/day)
Residential consumption during the assessment period/population served/assessment period

CI72 – Consumption per service connection (m^3/connection/year)
(Authorised consumption – exported water)/total number of *service connections* Variable ratio: (A14 – A5 – A7)/C24

Peak factors of supplied and exported water

CI73 – Monthly peak factor of supplied and exported water (-)
Actual monthly peak supplied and exported water × 12/supplied and exported water during the year (peak month/annual average)

CI74 – Daily peak factor of supplied and exported water (-)
Actual daily peak supplied and exported water × 365/supplied and exported water (peak day/annual average)

CI75 – Hourly peak factor of supplied and exported water (-)
Actual hourly peak supplied and exported water × 24 × 365/supplied and exported water (peak hour/annual average)

CI76 – Network delivery rate (m^3/km/year)
Authorised consumption (including exported water)/total mains length Variable ratio: A14/C8

CONTEXT INFORMATION

4.6 Demography and economics

CI77 – Population (No.)
Resident population within the supply area Also used as variable E5

CI78 – Population density (persons/km^2)
resident population/area under the responsibility of the *water utility*

CI79 – Household occupancy (persons/house)
resident population/total number of dwelling units (houses + apartments)

Population growth rate

CI80 – Current population growth rate (% per year)
Population variation during the last ten years/population in first year of this period/10
In fast growing areas a 5-year period should be adopted

CI81 – Forecasted population growth rate (% per year)
Forecasted average annual population growth rate for the future ten years

CI82 – Gross National Product per capita (EUR per capita/year)
Gross National Product/total country population

CI83 – Inflation rate (% per year)
Annual percentage change in consumer price index in the country.

EUR exchange rate

CI84 – Annual average EUR exchange rate (local currency/EUR)
Exchange ratio of the local currency for the EUR, assessed as the mean of the values at the end of each month, for the reference year.

CI85 – EUR exchange rate at the end of the assessment period (local currency/EUR)
Exchange ratio of the local currency for the EUR at the end of the assessment period.

4.7 Environment

These statistics relate to the area of service.

Annual rainfall

CI86 – Annual average rainfall (mm/year)
Annual average rainfall (average for the past 30 years)

CI87 – Annual maximum rainfall (mm/year)
Annual maximum rainfall assessed as the annual maxima of the last 30 years

CI88 – Annual minimum rainfall (mm/year)
Annual minimum rainfall assessed as the annual minima of the last 30 years

Air temperature

CI89 – Daily average air temperature (°C)
Average daily air temperature of the year (averages for the past 30 years)

CONTEXT INFORMATION

CI90 – Daily maximum air temperature (°C)
Average air temperature for the hottest day of the year (averages for the past 30 years)

CI91 – Daily minimum air temperature (°C)
Average air temperature for the coldest day of the year (averages for the past 30 years)

Topography

CI92 – Source average elevation (m)
Weighted average elevation above sea level of water sources, including import delivery points, using source production as weighing factors

CI93 – Maximum delivery elevation (m)
Maximum elevation above sea level at the water delivery points of water distribution area

CI94 – Minimum delivery elevation (m)
Minimum elevation above sea level at the water delivery points of water distribution area

Raw water quality – Source types

CI95 – Upland surface water sources (%)
Annual abstraction of *upland surface water*/total annual abstraction × 100

CI96 – Lowland surface water sources (%)
Annual abstraction of *lowland surface water*/total annual abstraction × 100

CI97 – Natural springs and wetlands sources (%)
Annual abstraction of *natural springs and wetlands water*/total annual abstraction × 100

CI98 – Well water sources (%)
Annual abstraction of *well water*/total annual abstraction × 100

CI99 – Borehole water sources (%)
Annual abstraction of *borehole water*/total annual abstraction × 100

CI100 – Saline and brackish water sources (%)
Annual abstraction of *saline and brackish water*/total annual abstraction × 100

5. UNCERTAINTIES AND UNCERTAINTY PROPAGATION

Accuracy, reliability and uncertainty

This section summarises the basis of the procedure that can be implemented to assess the PI accuracy bands. In I-6, the relevance of knowing the confidence grade of the input data are highlighted, and the concepts of reliability of the sources and accuracy of the data element are presented:

- Accuracy: The closeness of observations, computations or estimates to the true value as accepted as being true. Accuracy relates to the exactness of the result, and is distinguished from precision which relates to the exactness of the operation by which the result was obtained.
- Reliability (of the source): The extent to which data source yields consistent, stable, and uniform results over repeated observations or measurements under the same conditions each time.

When data are algebraically operated to generate a PI, the confidence of the result depends on the confidence of the input data. It does not seem very relevant or meaningful to retain the concept of reliability (of the source) for a value that is calculated from a combination of sources. However, it is deemed crucial to estimate how the uncertainty of the input data affects the uncertainty of the result.

According to the International Vocabulary of Basic and General Terms in Metrology uncertainty (of measurement) is a parameter, associated with the result of a measurement, that characterises the dispersion of the values that could reasonably be attributed to the measurand.

If the data provider indicates that the value a given data element is $A \pm U_A$, this means that he/she is confident that the true value is within a distance of U_A from A. U_A is designated (expanded) uncertaintainty. It is also usual to express uncertainty as a relative value, as a percentage of the mean of the variable it refers to. The term "is confident" may mean different things to different people. Statistically, it should be expressed in terms of a confidence level (i.e., the probability that the true value is in fact within the interval specified). A confidence interval of 95%, for instance, corresponds to the assumption that the data provider implicitly accepts that the probability of the true value being out of the band $[A - U_A; A + U_A]$ is not higher than 5% (one in twenty).

Uncertainty is generally caused by a broad combination of factors with randomly distributed effects. If this is the case, the difference between the true value and its measures is a variable with a normal distribution. Uncertainty can be measured in terms of the corresponding standard deviation (σ). In practice, the most common measures of uncertainty are:

$$U = 1\ \sigma \text{ (corresponding to a confidence level of ~68\%)}$$

or

$$U = 2\ \sigma \text{ (corresponding to a confidence level of ~95\%)}$$

or

$$U = 3\ \sigma \text{ (corresponding to a confidence level of ~99\%)}$$

Uncertainty is generally expressed in relative terms ($u_A = U_A/A$).

It is recommended that input data accuracy is specified in terms of an accuracy band, expressed in terms of relative uncertainty:

Accuracy band	Associated relative errors
0–5%	Better than or equal to ±5%
5–20%	Worse than ±5%, but better than or equal to +/–20%
20–50%	Worse than ±20%, but better than or equal to +/–50%
>50	Worse than ±50%

To assess the accuracy band of the PI, there is the need to use the concept of uncertainty.

Law of propagation of uncertainty

PI result from a combination of the four basic algebraic operations: sums, subtractions, products and divisions. Let us see how uncertainty is propagated in these simple cases, before looking at a more general expression.

Let us suppose that two measured quantities A and B have uncertainties U_A and U_B. We would report ($A \pm U_A$), and ($B \pm U_B$). From the measured quantities a new quantity, C, is calculated from A and B. What is the uncertainty, U_Z, in Z? For the purposes of this manual we will use a simplified version of the proper statistical treatment. Uncertainty propagation is described in detail in the ISO Guide to the Expression of Uncertainty in Measurement correlation coefficients[13].

(a) Addition and Subtraction: C = A + B + … or C = A − B − … or combinations of both:

The uncertainty of C is $U_C = \sqrt{(U_A)^2 + (U_B)^2 + \cdots}$.

(b) Multiplication and Division: C = A × B × … or C = A/B/… or combinations of both:

The relative uncertainty of C is $u_C = \sqrt{(u_A)^2 + (u_B)^2 + \cdots}$.

In this case it was implicitly assumed (for ease of understanding) that $U_w = 1\ \sigma_w$ (corresponding to a 68% confidence interval).

(c) General rule, for a quantity $z = f(w_1, w_2, \ldots, w_N)$ that depends on N input quantities $w_1, w_2, \ldots, w_N$, when $w_1, w_2, \ldots, w_N$ are mutually independent:

The average deviation of z is

$$U_z = \sqrt{\left|\frac{\partial f}{\partial w}\right|^2 (U_w)^2 + \left|\frac{\partial f}{\partial x}\right|^2 (U_x)^2 \left|\frac{\partial f}{\partial y}\right|^2 (U_y)^2 + \ldots}$$

(d) General rule, for a quantity $z = f(w_1, w_2, \ldots, w_N)$ that depends on N input quantities $w_1, w_2, \ldots, w_N$:

In this case the full general expression of uncertainty propagation is adopted.

13 Guide to the expression of uncertainty of uncertainty in measurement, International Organization for Standardization, Geneva (1993).

Assuming U = 1 σ, the uncertainty of z is

$$\sigma_z^2 = \sum_{i=1}^{N}\left(\frac{\partial f}{\partial w_i}\right)^2 \sigma_i^2 + 2\sum_{i=1}^{N} j \sum_{j=i+1}^{N} \frac{\partial f}{\partial w_i}\frac{\partial f}{\partial w_j}\rho_{ij}$$

where σ_z^2 is the variance of w_i and ρ_{ij} is the correlation coefficient of the covariance of w_i and w_j. If the input quantities are independent (as is often the case), then the covariance is zero and the second term of the above equation vanishes. The above equation was traditionally called the "general law of error propagation," but this equation actually shows how the uncertainties (not the errors) of the input quantities combine.

IWA Publishing's authorised EU representative for General Product Safety Regulations is Diane D'Arras, 15 rue Duret, 75116 Paris, France, e-mail: safety@iwap.co.uk.

Printed and bound by CPI Group (UK) Ltd, Croydon, CR0 4YY
27/03/2026
02079975-0003